Merchant Shi

Merchant Ship Types provides a broad and detailed introduction to the classifications and main categories of merchant vessels for students and cadets. It introduces the concept of ship classification by usage, cargo type, and size, and shows how the various size categories affect which ports and channels the types of vessels are permitted to enter. Detailed outlines of each major vessel category are provided, including:

- Feeder ship;
- General cargo vessels;
- Container ships;
- Tankers;
- Dry bulk carriers;
- Multi-purpose vessels;
- Reefer ships;
- Roll-on/roll-off vessels.

The book also explains where these are permitted to operate, the type of cargoes carried, and specific safety or risk factors associated with the vessel class, as well as their main characteristics. Relevant case studies are presented.

The textbook is ideal for merchant navy cadets at HNC, HND, and foundation degree level in both the deck and engineering branches, and serves as a general reference for insurance, law, logistics, offshore, and fisheries.

Merchant Ship Types

Alexander Arnfinn Olsen

LONDON AND NEW YORK

Cover image: Shutterstock

First published 2023
by Routledge
4 Park Square, Milton Park, Abingdon, Oxon OX14 4RN

and by Routledge
605 Third Avenue, New York, NY 10158

Routledge is an imprint of the Taylor & Francis Group, an informa business

British Library Cataloguing-in-Publication Data
A catalogue record for this book is available from the British Library

ISBN: 978-1-032-37876-3 (hbk)
ISBN: 978-1-032-37875-6 (pbk)
ISBN: 978-1-003-34236-6 (ebk)

DOI: 10.1201/9781003342366

Typeset in Sabon
by SPi Technologies India Pvt Ltd (Straive)

Dedicated to Captain Kåre B. Olsen
For whom there was nothing more enticing,
disenchanting, and enslaving than the life at sea
3 April 1940–22 March 2020

Contents

List of figures

List of tables

Preface

A cargo ship or freighter is a merchant ship that carries cargo, goods, and materials from one port to another. Thousands of cargo carriers ply the world's seas and oceans each year, managing the bulk of international trade. Cargo ships are usually specially designed for the task and are often being equipped with cranes and other mechanisms to load and unload and come in all shapes and sizes. Today, they are always built of welded steel, and with some exceptions, have a life expectancy of between 25 and 30 years before being scrapped. The words cargo and freight have become interchangeable in casual usage. Technically, 'cargo' refers to the goods carried by the ship for hire, while 'freight' refers to the act of carrying of said cargo, but the terms have been used interchangeably for centuries. The modern ocean shipping industry is divided into two sectors:

- Liner business: typically (but not exclusively) container vessels (wherein 'general cargo' is carried in 20 or 40-foot containers), operating as 'common carriers', calling at a regularly published schedule of ports. A common carrier refers to a regulated service where any member of the public may book cargo for shipment, according to long-established and internationally agreed rules
- Tramp-tanker business: this is business arranged between the shipper and receiver and facilitated by the vessel owners or operators, who offer their vessels for hire to carry bulk (dry or liquid) or break-bulk (cargoes with individually managed pieces) to any suitable port(s) in the world, according to a specifically drawn contract, called a charter party.

Larger cargo ships are operated by shipping lines: companies that specialise in the handling of cargo in general. Smaller vessels, such as coasters, are often owned by their operators. Ships are classified in numerous ways. The purpose of classifying ships is to differentiate between one type of ships and another. Cargo ships/freighters can be divided into seven groups, according to the type of cargo they carry. These groups are as follows:

- Feeder ship;
- General cargo vessels;
- Container ships;
- Tankers;
- Dry bulk carriers;
- Multi-purpose vessels;
- Reefer ships;
- Roll-on/roll-off vessels.

General cargo vessels carry packaged items such as chemicals, foods, furniture, machinery, motor- and military vehicles, footwear, and garments. Container ships (sometimes spelled containerships) are cargo ships that carry all their load in truck-size intermodal containers, using a technique called containerisation. They are a common means of commercial intermodal freight transport and now carry most seagoing non-bulk cargo. Container ship capacity is measured in 20-foot equivalent units (TEU). Tankers carry petroleum products or other liquid cargo. Dry bulk carriers carry coal, grain, ore, and other comparable products in a loose form. Multi-purpose vessels, as the name suggests, carry different classes of cargo – for example liquid, and general cargo – at the same time. A Reefer, reefer ship (or refrigerated), is specifically designed and used for shipping perishable commodities, which require temperature-controlled goods such as fruits, meat, fish, vegetables, dairy products, and other foodstuffs. Roll on, roll off (RORO or ro-ro) ships are designed to carry wheeled cargo, such as cars, trucks, semi-trailer trucks, trailers, and railroad cars, which are driven on and off the ship on their own wheels. Specialised types of cargo vessels include container ships and bulk carriers (technically tankers of all sizes are cargo ships, although they are routinely thought of as a separate category.

Cargo ships fall into two further categories that reflect the services they offer to industry: liner and tramp services. Those on a fixed published schedule and fixed tariff rates are cargo liners. Tramp ships do not have fixed schedules. Users charter them to haul loads. The smaller shipping companies and private individuals operate tramp ships. Cargo liners run on fixed schedules published by the shipping companies. Each trip a liner takes is called a voyage. Liners mostly carry general cargo; however, some cargo liners may also carry passenger. A cargo liner that carries 12 or more passengers is called a combination or a passenger-run-cargo line. Cargo ships are categorised partly by cargo capacity, partly by weight (deadweight tonnage (dwt), and partly by dimensions. Maximum dimensions such as length and width (beam) limit the canal locks a ship can fit in, water depth (draught) is a limitation for canals, shallow straits or harbours and height is a limitation to pass under bridges. The most common categories include *dry cargo ships*, which includes the following:

- Small handy size, carriers of 20,000–28,000 dwt;
- Seawaymax, 28,000 dwt, the largest vessel that can traverse the St. Lawrence Seaway. These are vessels less than 225.6 m (740 ft) in length, 23.8 m (78 ft) beam, and have a draught less than 8.08 m (26.51 ft) and a height above the waterline of no more than 35.5 m (116 ft);
- Handy size, carriers of 28,000–40,000 dwt;
- Handymax, carriers of 40,000–50,000 dwt;
- Panamax, the largest size that can traverse the original locks of the Panama Canal, a 294.13 m (965.0 ft) length, a 32.2 m (106 ft) beam, and a 12.04 m (39.5 ft) draught as well as a height limit of 57.91 m (190.0 ft). Limited to 52,000 dwt loaded, 80,000 dwt empty;
- Neopanamax, upgraded Panama locks with 366 m (1,201 ft) length, 55 m (180 ft) beam, 18 m (59 ft) draft, 120,000 dwt;
- Capesize, vessels larger than Suezmax and Neopanamax, and must traverse the Cape of Good Hope and Cape Horn to travel between oceans;
- Chinamax, carriers of 380,000–400,000 dwt up to 24 m (79 ft) draught, 65 m (213 ft) beam and 360 m (1,180 ft) length; these dimensions are limited by port infrastructure in China.

Wet cargo ships include the following:

- Aframax, oil tankers between 75,000 and 115,000 dwt. This is the largest size defined by the average freight rate assessment (AFRA) scheme.
- Q-Flex or Q-Max liquefied natural gas carrier for Qatar exports. A ship of Q-Max size is 345 m (1,132 ft) long and has a beam of 53.8 m (177 ft) and 34.7 m (114 ft) high, with a shallow draught of approximately 12 m (39 ft).
- Suezmax, typically ships of about 160,000 dwt, maximum dimensions are a beam of 77.5 m (254 ft), a draught of 20.1 m (66 ft) as well as a height limit of 68 m (223 ft) can traverse the Suez Canal
- VLCC (very large crude carrier), super tankers between 150,000 and 320,000 dwt.
- Malaccamax, ships with a draught less than 20.5 m (67.3 ft) that can traverse the Strait of Malacca, typically 300,000 dwt.
- ULCC (ultra large crude carrier), enormous super tankers between 320,000 and 550,000 dwt.

The TI-class super tanker is an ultra large crude carrier, with a draught that is deeper than the Suezmax, Malaccamax, and New Panamax classes. This causes Atlantic/Pacific routes to be exceptionally long, such as the long voyages south of Cape of Good Hope or south of Cape Horn to transit between the Atlantic and Pacific Oceans. Lake freighters, built for the Great Lakes in North America, differ in design from sea water–going ships because of the difference in wave size and frequency in the lakes. A number of these ships are larger than Seawaymax and cannot leave the lakes and pass to the Atlantic Ocean, as they do not fit the locks on the St. Lawrence Seaway.

A ship prefix is a combination of letters, usually, abbreviations, used in front of the name of a civilian or naval ship that has historically served numerous purposes, such as identifying the vessel's mode of propulsion, purpose, or ownership/nationality. In the modern environment, prefixes are used inconsistently in civilian service, whereas in government service, the vessels prefix is seldom missing due to government regulations dictating a certain prefix be present. Today, the customary practice is to use a single prefix for all warships of a nation's navy, and other prefixes for auxiliaries and ships of allied services, such as coast guards. For example, the modern navy of Japan adopts the prefix 'JS' – Japanese Ship. However, not all navies use prefixes; this includes the significant navies of China, France [though France has a NATO designation for its naval prefix, which is FS (French Ship)], Russia, and Spain, which like France uses the NATO designation SPS (Spanish Ship) if needed. Historically, prefixes for civilian vessels often identified the vessel's mode of propulsion, such as 'SS' (steam ship), 'MV' (motor vessel), or 'PS' (paddle steamer). Alternatively, they might have reflected a vessel's purpose, for example 'RMS' (Royal Mail Ship), or 'RV' (Research Vessel). These days, general civilian prefixes are used inconsistently, and frequently not at all. In terms of abbreviations that may reflect a vessel's purpose or function, technology has introduced a broad variety of differently named vessels onto the world's oceans, such as 'LPGC' (liquified petroleum gas carrier), or 'TB' (tugboat), or 'DB' (derrick barge). In many cases though, these abbreviations are used for purely formal, legal identification and are not used colloquially or in the daily working environment.

In terms of vessels used by a nations' armed services, prefixes not only primarily reflect ownership – but may also indicate a vessel's type or purpose as a subset. Historically, the most significant navy was Britain's Royal Navy, which has traditionally used the prefix 'HMS', standing for 'His / Her Majesty's Ship'. The Royal Navy also adopted nomenclature that reflected a vessel's type or purpose, for example HM Sloop. Commonwealth navies adopted a variation, with, for example, HMAS, HMCS, and HMNZS pertaining to Australia, Canada, and New Zealand, respectively. In the early days of the United States Navy, abbreviations often included the type of vessel, for instance 'USF' (United States Frigate), but this method was abandoned by President Theodore Roosevelt's Executive Order No. 549 of 1907, which made 'United States Ship' (USS) the standard signifier for USN ships on active commissioned service. In the United States Navy, that prefix officially only applies while the ship is in active commission, with only the name used before or after a period of commission and for all vessels 'in service' rather than commissioned status. However, not all navies use prefixes; this includes the significant navies of China, France, and Russia. From the 20th century onwards, most navies identify ships by letters or hull numbers (pennant numbers) or a combination of both. These identification codes were, and still are, painted on the side of the ship. Each navy has its own system: the United States Navy uses hull classification symbols, whereas the Royal Navy and most other European and Commonwealth navies use pennant numbers.

In this book, we will explore each of the main ship categories of merchant vessels in greater detail.

Acknowledgements

I would like to thank the many individuals who have provided me with assistance and material during the writing of this book, with particular thanks to Captain Reuben Lanfranco, Tony Moore, Aimee Wragg, and Divya Muthu. Also, to the many colleagues and friends who have answered numerous queries and added their wealth of experience, I extend my grateful thanks and gratitude.

Alexander Arnfinn Olsen
Southampton, October 2022

Abbreviations, glossary and terms used

AC	Alternating Current
ACP	Panama Canal Authority
AFRA	Average Freight Rate Assessment
AHTS	Anchor Handling Tug Supply
ANL	Australian National Lines
ATB	Articulated Tug and Barge
BIBO	Bulk In, Bag Out
BLEVE	Boiling Liquid Expanding Vapour Explosions
CDC	US Centres for Disease Control
CNG	Compressed Natural Gas
COA	Contract of Affreightment
COSCO	China Oriental Shipping Company
CS	Cable Ship
CV	Crane Vessel
DC	Direct Current
DFDE	Dual Fuel Diesel Electric
DP	Dynamic Positioning
DSME	Daewoo Shipbuilding & Marine Engineering
DSV	Diving Support Vessel
DWT	Deadweight
EEZ	Exclusive Economic Zone
EPS	Expanded Polystyrene
ERRV	Emergency Response and Rescue Vessel
ETV	Emergency Towing Vessel
FAO	Food and Agriculture Organisation
FC	Fully Cellular
FEMA	US Federal Emergency Management Agency
FLNG	Floating Liquefied Natural Gas
FPSO	Floating Production Storage and Offloading
FRV	Fisheries Research Vessel
FSO	Floating Storage and Offloading
FSRU	Floating Storage and Regasification Unit
FT	Foot / Feet

GARP	Global Atmospheric Research Programme
HAPAG	Hamburg-Amerikanische Packetfahrt-Actien-Gesellschaft
HFO	Heavy Fuel Oil
IACS	International Association of Classification Societies
ICAO	International Civil Aviation Organisation
ICBC	Industrial and Commercial Bank of China
IGY	International Geophysical Year
IMO	International Maritime Organisation
ILO	International Labour Organisation
IN	Inches
ISO	International Standards Organisation
ISPS	International Ship and Port Security Code
KG	Kilogrammes
KM/H	Kilometres Per Hour
KTS	Knots
LCE	Linear Cable Engine
LCTC	Large Car and Truck Carrier
LDT	Light Displacement Tonnage
LIMS	Lanes In Metres
LNG	Liquefied Natural Gas
LOA	Length Overall
LOLO	Lift On, Lift Off
LPG	Liquefied Petroleum Gas
LTBP	London Tanker Brokers' Panel
M	Metres
MARPOL	International Convention for the Prevention of Pollution from Ships
MLWS	Mean Low Water Springs
MPH	Miles Per Hour
MS	Motor Ship
MT	Motor Tanker
MV	Motor Vessel
NATO	North Atlantic Treaty Organisation
NAVARMA	Navigazione Arcipelago Maddalenino
NJ	New Jersey (US)
NOAAS	National Oceanic and Atmospheric Administration (US)
NY	New York (US)
NZALC	New Zealand and Australian Land Company
OBO	Oil Bulk Ore Carrier
OCV	Offshore Construction Vessel
OSV	Offshore Supply Vessel
PCC	Pure Car Carrier
PCTC	Pure Car and Truck Carrier
PIDN	Port Inland Distribution Network
PLEM	Pipeline End Manifold

PLV	Pipe Laying Vessel
PSV	Platform Supply Vessel
RAS	Replenishment at Sea
RFA	Royal Fleet Auxiliary
RMS	Royal Mail Ship
ROPAX	Roll-on/roll-off Passenger Ship
RORO	Roll-on/roll-off
ROV	Remotely Operated Vehicle
RRS	Royal Research Ship (UK)
RSHI	Jiangsu Rongsheng Heavy Industries
RV	Research Vessel
SOLAS	International Convention for the Safety of Life at Sea
SS	Steam Ship / Sailing Ship
STOVL	Short Take-Off and Vertical Landing
TEU	Twenty-Foot Equivalent
TFDE	Tri Fuel Diesel Electric
TLP	Tension Leg Platform
UK	United Kingdom
ULBC	Ultra Large Bulk Carrier
ULCC	Ultra Large Crude Carrier
ULOC	Ultra Large Ore Carrier
UN	United Nations
UNCTAD	United Nations Council on Trade and Development
US	United States
USCG	United States Coastguard
USCGC	United States Coastguard Cutter
USS	United States Ship
UTC	Universal Coordinated Time
VHSS	Vereinigung Hamburger Schiffsmakler und Schiffsagenten e. V.
VIP	Vacuum Insulated Panels
VLBC	Very Large Bulk Carrier
VLCC	Very Large Crude Carrier
VLCS	Very Large Container Ship
VLOC	Very Large Ore Carrier
WMO	World Meteorological Organisation
WSV	Well Stimulation Vessel

Introduction

Merchant ships are broadly classified based on their size and area of operation. The classification of a ship is determined at the design stage and includes the anticipated routes of operation and the purpose of the ship. The ship's dimensions play an important role in determining the areas of operation of any type of merchant vessel. A variety of parameters such as draught, beam, length overall, gross tonnage, deadweight (dwt) tonnage, and so forth are taken into consideration when designing and constructing the merchant ship. For example, when designing a ship that will pass through the Suez Canal, the dimensions of the ship will be decided in such a way that the ship is able to smoothly transit through the narrowest and shallowest parts of the canal, in both fully loaded and unloaded conditions. This is determined by the ship's load carrying capacity. There are many different classifications and sub-classifications of merchant vessels; therefore, this book will focus primarily on the main categories that readers are likely to come across. There are 10 main classifications of ships based on their size. We use size as the primary determining feature as this is the most obvious characteristic of the ship. Also, as we mentioned before, there are hundreds of types of ships from tugboats to oil tankers. We will start by first examining the main classifications of ships by size, followed by their category. By category, we mean the dominant type of cargo carried, such as containers, passengers, refrigerated cargoes, and vehicles.

SHIP CLASSIFICATIONS BY SIZE

In terms of classifying ships by their size, there are 10 categories, which are most frequently used. These are the Panamax and New Panamax, Aframax, Chinamax, Handymax, Suezmax, Capesize, Q-Max, Malaccamax, Very Large (VLCC) and Ultra Large Crude Carriers (ULCC), and Seawaymax. As you might have guessed from the name, these classification types are named after important seaways. This tells us that the ship in that category is of the largest size that is permitted to enter that area.

DOI: 10.1201/9781003342366-1

Aframax

The term Aframax is usually used for medium-sized oil tankers with an approximate weight of 120,000 dead weight metric tonnes (dwt). These ship sizes gain their title from the average freight rate assessment (AFRA) schematic devised by the shipping conglomerate Shell in the mid-1950s. Applied to oil tankers, Aframax vessels can be loaded with over seven million barrels of crude oil. Aframax tankers ply in areas that have limited port facilities or lack large ports to accommodate larger oil carriers. The beam of the vessel is restricted to 32.3 m (106 ft).

Baltimax

Baltimax is a naval architecture term for the largest ship measurements capable of entering and leaving the Baltic Sea in a laden condition. It is the Great Belt route that allows the largest ships. The limit is a draught of 15.4 m (50.52 ft) and a height above the waterline of 65 m (541.33 ft), which is limited by the clearance of the east bridge of the Great Belt Fixed Link. Baltimax vessels typically have a maximum length overall of 240 m (787 ft) and a beam of 42 m (137 ft). This gives a weight of around 100,000 metric tonnes. Nevertheless, there are also certain larger ship types plying the Baltic Sea. In particular, the B-Max crude oil tanker has a gross dwt of 205,000 metric tonnes, a length overall of 325 m (1066 ft), and a beam of 68 m (223 ft). The largest container ship to traverse the Baltic Sea is the *'Mærsk Triple E class'* container ship, which has a length overall of 400 m (1,312 ft) and 165,000 metric tonnes dwt. The Öresund permits a maximum draught of only 8.0 m (26.24 ft) and is effectively closed to all larger vessels. The Nord-Ostsee-Kanal also has a shallow draught of 9.5 m (31.16 ft). Furthermore, many of the ports located on the Baltic Sea limit the ship size. The iron ore ports of Luleå (Sweden) (11.0 m, 36.08 ft), for example, were deepened to 13 m (42.65 ft) and Kemi (Finland) has a maximum allowable draught of 10.0 m (32.80 ft). Even the large port of Klaipėda (Lithuania) only has a permitted draught of 13.8 m (45.27 ft), which is less than most of the Baltimax vessels. The largest port on the Baltic Sea is Primorsk (Russia), which has a maximum permissible draught of 15 m (49.21 ft), which can accommodate the Baltimax provided they are not fully laden. Alternatively, the Northern Port in Gdańsk (Poland) can accommodate ships with a maximum draught of 15 m (49.21 ft) and a total tonnage of 300,000 dwt.

Capesize

Capesize ships are the largest dry cargo ships. They are too large to transit the Suez Canal (Suezmax limits) or Panama Canal (Neopanamax limits), and so must pass either the Cape Agulhas or Cape Horn to traverse between oceans. Ships in this class are bulk carriers, usually transporting coal, ore, and other commodity raw materials. The term Capesize is not applied to tankers. The average size of a Capesize bulker is around 156,000 dwt, although larger ships (normally dedicated to ore transportation) have been built, up to 400,000 dwt. The large dimensions and deep draughts of such vessels mean that only the largest deep-water terminals can accommodate them.

Chinamax

Chinamax is a standard of ship measurements that allow conforming ships to use various harbours when fully laden; the maximum size of such ships is a draught of 24 m (79 ft), a beam of 65 m (213 ft), and a length overall of 360 m (1,180 ft). An example of ships of this size is the Valemax bulk carriers (discussed below). The standard was originally developed to carry exceptionally large loads of iron ore to China from Brazilian port facilities operated by the mineral mining firm Vale. Correspondingly, ports and other infrastructure that are 'Chinamax-compatible' are those at which such ships can readily dock. Unlike Suezmax and Panamax, Chinamax is not determined by locks, channels, or bridges. The Chinamax standard is aimed at port provisions and the name is derived from the massive dry-bulk (ore) shipments that China receives from around the globe. In container shipping, recent classes intended for trade with China have all focused on a ~400 m (~1,312 ft) length, which deep water container terminals can cater for.

Handysize

Handysize is a naval architecture term used for smaller bulk carriers and oil tankers with a maximum dwt of up to 50,000 metric tonnes, although there is no official definition in terms of exact tonnages. Handysize may also sometimes refer to a span of up to 60,000 metric tonnes, with vessels above 35,000 metric tonnes referred to as Handymax or Supramax. Their small size allows Handysize vessels to enter smaller ports to pick up cargoes, and because in most cases, they are 'geared' – that is fitted with cranes – they can often load and discharge cargoes at ports, which lack cranes or other integrated cargo handling systems. Compared with larger bulk carriers, Handysize carry a wider variety of cargo types. These include steel products, grain, metal ores, phosphate, cement, wood logs, woodchips, and other types of so-called 'break-bulk cargo'. They are numerically the most common size of bulk carrier, with 2,000 units in service totalling about 43 million metric tonnes. Handysize bulkers are built by shipyards in Japan, South Korea, China, Vietnam, the Philippines and India, though a few other countries also have the capacity to build this class of ship. The most common industry-standard specification is that the Handysize bulker is about 32,000 metric tonnes dwt on a summer draught of about 10 m (33ft) and features five cargo holds with hydraulically operated hatch covers, with

43 metric tonne cranes for cargo handling. Some Handysize are also fitted with stanchions to enable logs to be loaded in stacks on deck. Such vessels are often referred to as 'handy loggers'. Despite multiple recent orders for new ships, the Handysize sector still has the highest average age profile of the major bulk carrier sectors.

Handymax

Handymax and Supramax are naval architecture terms for the larger bulk carriers in the Handysize class. Handysize class consists of Supramax (50,000–60,000 dwt), Handymax (40,000–50,000 dwt), and Handy (<40,000 dwt). The ships are used for less voluminous cargoes, and different cargoes can be carried in different holds. Larger capacities for dry bulk include Panamax, Capesize and Very Large Ore Carriers (VLOC), and Chinamax. Handymax ships typically have a length overall of 150–200 m (492–656 ft), though certain bulk terminal restrictions, such as those in Japan, mean that many Handymax ships are just under 190 m (623 ft) in overall length. Modern Handymax and Supramax designs typically have a dwt of 52,000–58,000 metric tonnes, have five cargo holds and four cranes of around 30 metric tonnes working load, making it easier to use in ports with limited infrastructure. Average speeds depend on size and age. The cost of building a Handymax is driven by the laws of supply and demand. In early 2007, the cost of building a new Handymax was around US$20 million. As the global economy boomed, the cost doubled to over US$40 million, as demand for vessels of all sizes exceeded available yard capacity. Following the global economic crisis in 2009 and 2010, the cost tumbled to US$20 million. By 2018, the average price of a new Handymax vessel was £23 million.

Malaccamax

Malaccamax is a naval architecture term for the largest tonnage of ship capable of fitting through the 25m deep (82 ft) Strait of Malacca. Bulk carriers and super tankers have been built to this tonnage, and the term is chosen for very large crude carriers (VLCCs). These vessels can transport oil from the Persian Gulf and Saudi Arabia to China. A typical Malaccamax tanker has a maximum length of 333 m (1,093 ft), a beam of 60 m (197 ft), a draught of 20.5 m (67.3 ft), and a tonnage of 300,000 dwt. Any post-Malaccamax ship would need to use even longer alternate routes as traditional seaways such as the Sunda Strait, between the Indonesian islands of Java and Sumatra, are too shallow for large ships. Other routes would therefore be required, such as:

- The Lombok Strait [250 m (820 ft)], Dewakang Sill [680 m (2,230 ft)], Makassar Strait, then either east past Mindanao to the Philippine Sea or north through the Sibutu Passage and Mindoro Strait;

- Ombai Strait, Banda Sea, Lifamatola Strait [1,940 m (6,360 ft)] and between the Sula Islands and Obi Islands, and the Malacca Sea;
- Or around Australia.

Alternatively, artificially excavated new routes might also be a possibility, such as by deepening the Strait of Malacca, specifically at its minimum depth in the Singapore Strait, or the proposed Kra Canal, which would require extensive excavation.

Panamax and New Panamax

Panamax and New Panamax (or Neopanamax) are terms used for the size limits for ships travelling through the Panama Canal. The limits and requirements are published by the Panama Canal Authority (ACP) in a publication titled 'Vessel Requirements'. The requirements also provide notices for mariners concerning exceptional dry seasonal limits, propulsion, communications, and detailed ship design. The allowable size is limited by the width and length of the available lock chambers, by the depth of water in the canal, and by the height of the Bridge of the Americas. These dimensions give clear parameters for ships destined to traverse the Panama Canal and have influenced the design of cargo ships, naval vessels, and passenger ships (Table 1.1).

Panamax specifications have been in effect since the opening of the canal in 1914. In 2009, the ACP published the New Panamax specification, which came into effect when the Canal's third set of locks, larger than the original two, opened on 26 June 2016. Ships that do not fall within the Panamax sizes are called post-Panamax or super-Panamax. The increasing prevalence of vessels of the maximum size is a problem for the canal, as a Panamax ship is a tight fit that requires precise control of the vessel in the locks, resulting in longer lock time and requiring that these ships transit in daylight. Because the largest ships traveling in opposite directions cannot pass safely within the Culebra Cut, the canal effectively operates an alternating one-way system for these ships.

Table 1.1 Panamax vessel dimensions

	Locks	*Panamax*	*New locks*	*New Panamax*
Length	320.04 m (1050 ft)	294.13 m (965 ft)	427 m (1400 ft)	366 m (1,200 ft)
Width	33.53 m (110 ft)	32.31 m (106 ft)	55 m (180.5 ft)	49 m (160.7 ft)
Draught	12.56 m (41.2 ft)	12.04 m (39.5 ft)	18.3 m (60 ft)	15.2 m (49.9 ft)
TEU		5,000		12,000

Ship dimensions

Panamax is determined principally by the dimensions of the Canal's original lock chambers, each of which is 33.53 m (110 ft) wide, 320.04 m (1,050 ft) long, and 12.56 m (41.2 ft) deep. The usable length of each lock chamber is 304.8 m (1,000 ft). The available water depth in the lock chambers varies, but the shallowest depth is at the south sill of the Pedro Miguel Locks and is 12.56 m (41.2 ft) at a Miraflores Lake level of 16.61 m (54.06 ft). The clearance under the Bridge of the Americas at Balboa is the limiting factor on a vessel's overall height for both Panamax and Neopanamax ships; the exact figure depends on the water level. The maximum dimensions allowed for a ship transiting the canal using the original locks and the new locks (New Panamax) are discussed in the following sections.

Length

Overall (including protrusions): 289.56 m (950 ft) with the following exceptions:

- Container ships and passenger ships: 294.13 m (965 ft);
- Tug-barge combination, rigidly connected: 274.32 m (900 ft) overall;
- Other non-self-propelled vessels-tug combination: 259.08 m (850 ft) overall;
- New Panamax increases allowable length to 366 m (1,201 ft).

Beam (width)

The maximum permitted width over the outer surface of the shell plating is 32.31 m (106 ft) with a general exception of 32.61 m (107 ft) when the vessel's draught is less than 11.3 m (37 ft) in tropical fresh water. New Panamax vessels were originally allowed a beam of 49 m (161 ft) although this was extended in June 2018 to 51.25 m (168.14 ft).

Draught

The maximum allowable draught is 12.04 m (39.5 ft) in tropical freshwater. The name and definition of tropical freshwater is determined by ACP using the freshwater Lake Gatún as a reference, since this is the determination of the maximum draught. The salinity and temperature of water affects its density, and hence how deep a ship will float in the water. The physical limit is set by the lower (seaside) entrance of the Pedro Miguel locks. When the water level in Lake Gatún is low during an exceptionally dry season, the maximum permitted draft may be reduced. Such a restriction is published three weeks in advance, so ship loading plans can respond appropriately (Figure I.1).

Figure I.1 Panamax sized container vessel passing through the Panama Canal.

Height

Vessel height is limited to 57.91 m (190 ft) measured from the waterline to the vessel's highest point; the limit also relates to New Panamax vessels to pass under the Bridge of the Americas at Balboa harbour. The only exception is passage takes place at low water (MLWS) at Balboa, allowing vessels with a maximum height of 62.5 m (205 ft).

Cargo capacity

Panamax ships typically have a dwt of 65,000–80,000 metric tonnes when not passing through the Canal and a maximum dwt of 52,500 metric tonnes during transit. New Panamax ships can carry up to 120,000 metric tonnes dwt, which is equivalent to 5,000 TEU for Panamex vessels and 13,000 TEU for New Panamax vessels.

Records

The longest ship ever to transit the original locks was the *'San Juan Prospector'*, now known as the *'Marcona Prospector'*, an ore-bulk-oil carrier, with a length overall of 297 m (973 ft) and a beam of 32 m (106 ft). The widest ships to transit the Panama Canal were the four US Navy *'South Dakota class'* and *'Iowa class'* battleships, each having a maximum beam of 32.97 m (108.2 ft), leaving less than a 15 cm (6 in) margin of error between the ship side and the walls of the locks.

Expansion of the Panama Canal locks

As early as the 1930s, new locks were proposed for the Panama Canal to ease congestion and to allow larger ships to pass. The project was abandoned in 1942. On 22 October 2006, the ACP (with the support of the Electoral Tribunal) held a referendum for Panamanian citizens to vote on the Panama Canal expansion project. The expansion was approved by a wide margin, with support from about 78% of voters. Construction began in 2007, and after several delays, the new locks opened for commercial traffic on 26 June 2016. Several ports, including the ports of New York and New Jersey, Norfolk, and Baltimore, all on the East Coast of the United States (US), increased their depth to at least 15 m (50 ft) to accommodate the New Panamax ships. In 2015, the Port of Miami achieved the same in a project referred to as the 'Deep Dredge' and is now the closest United States deep-water port to the Panama Canal. In the United Kingdom (UK), the Port of Liverpool Authority commissioned a new container terminal, Liverpool 2, where ships can berth in the tidal river rather than in the enclosed docks, coinciding with the opening of the widened Panama Canal locks. In Halifax, Canada, a major expansion of the South End Container Terminal was completed in 2012, extending the pier and increasing the berth depth from 14.5 to 16 m (from 48 to 52 ft). In 2017, the Port Authority of New York and New Jersey raised the clearance of the Bayonne Bridge to 66 m (215 ft) at a cost of US$1.7 billion, to allow New Panamax ships to reach container port facilities at Port Newark–Elizabeth Marine Terminal. Previously, only GCT Bayonne, Global Container (New Jersey) could manage the New Panamax ships. In April 2012, a controversy erupted between the ports of Savannah, Georgia, and Charleston, South Carolina, over limited federal funding for dredging deepening projects. Both state and federal lawsuits filed by environmental groups in both states opposed the techniques planned to be used in dredging the Savannah River. Elsewhere in the United States, Jacksonville, Florida, has pursued its 'Mile Point' project with the aim of deepening the St. John's River to allow post-Panamax traffic; Mobile, Alabama, has completed the deepening of its harbour to 14 m (45 ft) for the same reason; and other ports seem likely to follow. Due to the expansion, demand for the 'old Panamax' ships has plummeted, resulting in ships being traded at scrap value. Some ships, as young as seven years, have been sold for scrap (Figure I.2).

Post-Panamax and Post-neo-panamax ships

Post-Panamax or Post-neo-panamax denotes ships that are larger than the Panamax designation and will not fit in the original canal locks. This includes super tankers and the largest modern container and passenger ships. The first post-Panamax ship was the *RMS Queen Mary*, launched in 1934, with a 39.56 m (118 ft) beam, as she was intended solely for

Figure I.2 Providence Bay passing through the Panama Canal.

North Atlantic passenger runs. When she was moved to Long Beach, California, as a tourist attraction in 1967, a lengthy voyage around the Cape Horn was required. The first post-Panamax warships were the Japanese *Yamato class* battleships, launched in 1940. Until World War II, the US Navy required that all their warships be capable of transiting the Panama Canal. The first US Navy warship design to exceed Panamax limits was the *Montana class* battleship, designed circa 1940, but never built. This limit was removed by the US Secretary of the Navy on 12 February 1940, with the (never-realised) prospect of a new set of 42.67 m (140 ft) wide locks to be built for the Canal. In the end, the *Essex class* aircraft carriers were designed with a folding deck-edge elevator to meet Panamax limits, although the limit did not apply to subsequent US aircraft carrier designs. The construction of the third set of larger locks has led to the creation of the Neo-panamax or New Panamax ship classification, based on the new locks' dimensions of 427 m (1,400 ft) in length, 55 m (180 ft) beam and 18.3 m (60.0 ft) depth. With the new locks, the Panama Canal is able to handle vessels with a length overall of 366 m (1,201 ft), 49 m (160.7 ft) beam although this was later increased by the ACP (effective 1 June 2018) to 51.25 m (168.14 ft) to accommodate ships with 20 rows of containers and 15.2 m (49.86 ft) draught and a cargo carrying capacity of up to 14,000 TEU. This compares favourably to the previous limits, which could only manage vessels up to about 5000 TEU.

Qatar Flex (Q-Flex)

The Q-Flex is a classification of membrane-type liquefied natural gas carrier. Q-Flex vessels are propelled by two slow speed diesel engines, which are claimed to be more efficient and environmentally friendly than traditional steam turbines. Q-Flex carriers are equipped with an on-board re-liquefaction system to handle boil-off gas, liquefy it and return the LNG to the cargo tanks. This on-board re-liquefaction system reduces LNG losses, which produces economic and environmental benefits. Q-Flex vessels are equipped with an on-board re-liquefaction system to handle boil-off gas, liquefy it and return the LNG to the cargo tanks. The on-board re-liquefaction system allows a reduction of LNG loss, which provides increased economic and environmental efficiencies. Overall, it is estimated that Q-Flex carriers operate with about 40% lower energy requirements and carbon emissions than conventional LNG carriers. In August 2008, 16 named Q-Flex LNG carriers were brought into commission: *Al Hamla*, *Al Gharrafa*, *Duhail*, *Al Ghariya*, *Al Aamriya*, *Murwab*, *Fraiha*, *Al Huwaila*, *Al Kharsaah*, *Al Shamal*, *Al Khuwair*, *Al Oraiq*, *Umm Al Amad*, *Al Thumama*, *Al Sahla*, and *Al Utouriya*. All these vessels are owned by holding companies established by the Qatar Gas Transport Company (Nakilat) and different shipping companies, including the Overseas Shipholding Group, Pronav, and Commerz Real, and are chartered to Qatar's LNG producers Qatargas and RasGas. The capacity of the Q-Flex vessel is between 165,000 and 216,000 m^3. Up until the entry into service of the Q-Max carrier, the Q-Flex was the world's largest LNG carrier type with a capacity of 1.5 times that of conventional LNG carriers (Figure I.3).

Qatar Max (Q-Max)

Like the Q-Flex, the Qatar Max or Q-Max is a classification of a membrane-type liquefied natural gas carrier and is the maximum size of the ship's ability to dock at the LNG terminals in Qatar. Ships of this type are the largest LNG carriers in the world. Q-Max vessels have a maximum length overall of 345 m (1,132 ft), a beam of 53.8 m (177 ft), a draught of approximately 12 m (39 ft), and a height above the water line of 34.7 m (114 ft). Q-Max vessels have an LNG capacity of 266,000 cu/m (9,400,000 cu/ft), equal to 161,994,000 cu/m (5.7208 × 109 cu/ft) of natural gas. These ships are propelled by two slow speed diesel engines burning heavy fuel oil (HFO), which are claimed to be more efficient and environmental-friendly than traditional steam turbines. In the event of engine failure, the failed engine can be decoupled allowing the ship to maintain a maximum speed of 14 knots. Q-Max vessels are equipped with an on-board re-liquefaction system to handle boil-off gas, liquefy it and return the LNG to the cargo tanks. The on-board re-liquefaction system allows a reduction of LNG loss, which provides increased economic and environmental efficiencies. Overall,

Figure I.3 LNG Carrier *Al Aamriya*, Milford Haven, Wales.

it is estimated that Q-Max carriers operate with about 40% lower energy requirements and carbon emissions than conventional LNG carriers. The quoted estimates do however ignore the additional fuel used to re-liquify boil off gas rather than burn the gas for fuel. The ships run on HFO; however, the *Rasheeda* was retrofitted with gas-burning capabilities in 2015. The first Q-Max LNG carrier was floated out of dry dock in November 2007. The naming ceremony was held on 11 July 2008 at Samsung Heavy Industries' shipyard on Geoje Island, South Korea. Known before its naming ceremony as *Hull 1675*, the ship was named *Mozah* by Sheikha Mozah Nasser al-Misnad. *Mozah* was delivered on 29 September 2008. Classed by Lloyd's Register, *Mozah* completed her maiden voyage on 11 January 2009, when the tanker delivered 266,000 cu/m of LNG to the Port of Bilbao BBG Terminal. As of 2022, Q-Max LNG carriers are operated by STASCo (Shell International Trading and Shipping Company, London), which is a subsidiary of Royal Dutch Shell. The vessels themselves are owned by the Qatar Gas Transport Company (Nakilat) and chartered to Qatar's LNG producers Qatargas and RasGas. In total, contracts were signed for the construction of 14 Q-Max vessels: *Mozah*, *Al Mayeda*, *Mekaines, Al Mafyar*, *Umm Slal, Bu Samra*, *Al Ghuwairiya*, *Lijmiliya*, *Al Samriya*, *Al Dafna*, *Shagra*, *Zarga*, *Aamira*, and *Rasheeda*. All 14 Q-Max ships were delivered between 2008 and 2010.

Seawaymax

The term Seawaymax refers to vessels, which are the maximum size that can fit through the canal locks of the St. Lawrence Seaway, linking the inland Great Lakes of North America with the Atlantic Ocean. Seawaymax vessels have a maximum length overall of 230 m (740 ft), a beam of 24 m (78 ft), a draught of 8.1 m (26.51 ft), and a height above the waterline of 35.4 m (116 ft). Several lake freighters larger than this size cruise the Great Lakes but cannot pass through to the Atlantic Ocean. The size of the locks limits the size of the ships, which can pass, and consequently limits the size of the cargoes they can carry. The record tonnage for one vessel on the Seaway is 28,502 metric tonnes of iron ore, while the record through the larger locks of the Great Lakes Waterway is 72,351 metric tonnes. Most new lake vessels, however, are constructed to the Seawaymax limit to enhance vessel versatility.

Suezmax

The term Suezmax is used for the largest ships that can pass through the Suez Canal. The current channel depth of the canal allows for a maximum of 20.1 m (66 ft) of draught, meaning that few fully laden super tankers can fit through. This means that they must either unload part of their cargo to other ships (a process called 'transhipment') or pipe their cargo ashore to via pipeline to a terminal before passing through. The only other alternative is to bypass the Suez Canal completely and sail around the Cape Agulhas instead. To try and avoid this, the canal was deepened from 18 to 20 m (from 59 to 66 ft) in 2009. The typical dwt of a Suezmax ship is about 160,000 metric tonnes and has a beam of about 77.5 m (254.3 ft). Also of note is the maximum head room – or the 'air draught', which sets a limitation of 68 m (223.1 ft). This limitation is because of the Suez Canal Bridge, which rises some 70 m (230 ft) above the canal. The Suez Canal Authority produces tables of width and acceptable draught, which are regularly subject to change. From 2010, the wetted surface cross-sectional area of a ship was limited by 1006 m^2, which means 20.1 m (66 ft) of draught for ships with the beam no wider than 50.0 m (164.0 ft) or 12.2 m (40 ft) of draught for ships with a maximum allowed beam of 77.5 m (254 ft). Most Capesize vessels are too big to pass through the Suez Canal, and therefore need to travel the Cape route around the Cape of Good Hope and Cape Agulhas, though recent dredging operations mean that many Capesize vessels can now use the canal. Plans to deepen the draught to 21 m (70 ft) could lead to a redefinition of the Suezmax specification, as happened to the Panamax specification after the deepening and widening of the Panama Canal. As we discussed earlier, the Aframax is a freight rating, not a geographic routing limiter, for tankers with a capacity ranging between 80,000 and 120,000 metric tonnes dwt. Vessels longer than 400 m (1312 ft) need permission from the Suez Canal Authority to transit the canal. As of 2020, the largest container ships in service all have a length of (close to) 400 m (1,312 ft) and a beam and draught that just fits within the limits of the Canal. This issue was exemplified in 2021 when the ship *Ever Given* ran aground in the Suez Canal. This ship has a length overall of 399.9 m (1,312 ft) and a beam of 58.8 m (192.9 ft).

Valemax

Valemax ships are a fleet of VLOC owned or chartered by the Brazilian mining company Vale SA to carry iron ore from Brazil to European and Asian ports. With a capacity ranging from 380,000 to 400,000 metric tonnes dwt, the vessels meet the Chinamax standard of ship measurements for limits on draught and beam. Valemax ships are the largest bulk carriers ever constructed, when measured by dwt tonnage or length overall, and are amongst the longest ships of any type currently in service. The first Valemax vessel, *Vale Brasil*, was delivered in 2011. Initially, all 35 ships of the first series were expected to be in service by 2013, but the last ship was not delivered until September 2016. In late 2015 and early 2016, Chinese shipping companies ordered 30 more ships with deliveries arriving between 2018 and 2020. Three additional vessels were ordered by a Japanese shipping company, bringing the total number of Valemax vessels to 68 as of 2020. In 2008, Vale placed orders for twelve 400,000-tonne Valemax ships to be constructed by Jiangsu Rongcheng Heavy Industries (RSHI) in China and ordered seven more ships from the South Korean Daewoo Shipbuilding & Marine Engineering (DSME) in 2009. In addition, 16 more ships of equivalent size were ordered from Chinese and South Korean shipyards for other shipping companies and chartered to Vale under long-term contracts. The first vessel was delivered in 2011 and the last in 2016. The contract, worth US$1.6 billion, was the world's biggest single shipbuilding contract by dwt tonnage. The first Chinese-built Valemax vessel, *Vale China*, was launched at the Nantong shipyard on 9 July 2011 and was delivered on 25 November 2011. Although it was expected that the first Chinese-built Valemax vessel would call at a Chinese port on its maiden voyage, the ship was diverted to the new trans-shipment hub Vale, which was constructed in the Philippines. The second RSHI-built Valemax ship for Vale (*Vale Dongjiakou*) was delivered on 9 April 2012, the third (*Vale Dalian*) on 20 May, the fourth (*Vale Hebei*) on 28 September, the fifth (*Vale Shandong*) on 7 December 2012, the sixth (Vale Jiangsu) on 23 March 2013, the seventh (*Vale Caofeidian*) on 22 July 2013, the eighth (*Vale Lianyungang*) on 22 November 2013, the ninth (*Ore Majishan*; renamed before delivery) on 11 July 2014, the tenth (*Ore Tianjin*; renamed before delivery) on 18 October 2014, and the eleventh (*Ore Rizhao*; renamed before delivery) on 15 December 2014. The twelfth and last Valemax vessel of the original order by Vale, *Ore Ningbo* (renamed before delivery), was delivered on 23 January 2015. On 2 November 2008,

Oman Shipping Company signed a framework agreement with RSHI for the construction of four 400,000-tonne vessels to transport iron ore from Brazil to the Port of Sohar in Oman, where Vale was expected to open a steel plant. The shipbuilding contract, worth an estimated US$483 million, was signed in July 2009. Initially, the ships were to be named *Jazer*, *Yanqul*, *Al Kamil*, and *Wafi*, but instead were named *Vale Liwa*, *Vale Sohar*, *Vale Shinas*, and *Vale Saham*, respectively. The steel cutting ceremony for the first two vessels was held on 8 July 2010, which were launched on 19 March 2012. *Vale Liwa* entered service in August 2012, followed by *Vale Sohar* in September 2012, *Vale Saham* in January 2013, and *Vale Shinas* in March 2013. The ships received additional strengthening due to the *Vale Beijing* incident. The ships built for Oman Shipping Company were later removed from the Det Norske Veritas registry and moved to other classification societies such as American Bureau of Shipping and Lloyd's Register (Figure I.4).

The Chinese shipbuilder's ability to deliver any of the VLOC ordered by Vale in time was doubted before the first ship was even built. In May 2011, it was announced that only two or three Valemax vessels would be delivered from the Chinese shipyard in 2011 instead of the planned six due to delays in construction. In the end only, one ship (*Vale China*) was delivered before the end of the year. Furthermore, later reports claimed that the ships ordered by Vale had a capacity of only 380,000 metric tonnes even though, according to the Det Norske Veritas database entries, all Chinese-built ships have a dwt tonnage more than 400,000 metric tonnes, and in the past, Vale has referred to the ships ordered from Rongcheng as '400,000-tonne' vessels. The reduction in cargo capacity, at least on paper, may have been due to the

Figure I.4 Construction of the *Vale Dalian*.

reluctance of Chinese officials to accept the 400,000-tonne ships into Chinese ports. In April 2012, it was reported that Vale had refused delivery for three Valemax ships recently completed by Jiangsu Rongcheng Heavy Industries. This was seen as a move against the Chinese officials who had blocked the admittance of the 400,000-tonne ships to dock in Chinese ports. These reports however were refuted by RSHI, who called them 'inaccurate and unfounded'. On 26 October 2009, Vale ordered four Valemax vessels from the South Korean shipbuilder Daewoo Shipbuilding & Marine Engineering (DSME) for US$460 million. A further three ships were ordered from DSME in July 2010, bringing the total order to seven 400,000-metric tonne Valemax vessels. Despite receiving the order later than the Chinese shipyard, DSME launched the first *Valemax class* ore carrier, *Vale Brasil*, on 31 December 2010 and delivered the ship to Vale in March 2011; the *Vale Brasil* was followed by the *Vale Rio de Janeiro* on 22 September 2011, *Vale Italia* on 25 October 2011, *Vale Malaysia* on 27 March 2012, *Vale Carajas* on 29 May 2012 and *Vale Minas Gerais* on 13 July 2012. The last Valemax ship to be built by DSME, *Vale Korea*, was delivered on 9 April 2013. In addition to the ships Vale ordered for itself, more ships of a comparable size were ordered by other shipping companies and chartered to Vale under exclusive long-term contracts. Eight VLOC were ordered from the South Korean shipbuilder STX Offshore & Shipbuilding in Jinhae, South Korea (STX Jinhae), and Dalian, China (STX Dalian). The shipping company, STX Pan Ocean, signed a 25-year contract with Vale in 2009. The dwt tonnage of the Valemax vessels built by STX, 374,400 tons, was slightly smaller than that of the similar ships built by DSME and RSHI.

The first STX-built Valemax vessel, *Vale Beijing*, was delivered by STX Jinhae on 27 September 2011. Although another vessel was expected to take delivery later that year, only one ship was delivered. The second ship, *Vale Qingdao*, was also delivered by STX Jinhae on 13 April 2012, but the third and fourth ships, *Vale Espirito Santo* and *Vale Indonesia*, were built by STX Dalian and delivered on 17 September 2012 and 30 October 2012, respectively. The fifth ship, *Vale Fujiyama*, was again built by STX Jinhae and delivered on 26 November 2012. The sixth ship, *Vale Tubarao*, was delivered by STX Dalian on 30 January 2013. While both two remaining ships were supposed to be delivered by the end of 2013, only *Vale Maranhao* entered service on 29 August 2013. The last Valemax vessel to be built by STX, originally named *Vale Ponta da Madeira* but later referred to by its yard number *STX Dalian 1707*, was launched sometime in 2015. However, the vessel was never finished and instead was listed as 'for sale' in an unfinished state. In March 2016, it was reported that the last of the original 35 Valemax vessels to be built had been sold for just US$16.8 million. Initially anonymous, the buyer was later identified as Pan Ocean (formerly STX Pan Ocean), the shipping company who originally ordered the vessel. The unfinished vessel was completed by the Chinese shipyard Shanhaiguan Shipbuilding and named *Sea Ponta da Madeira*. On 30 April 2007,

Berge Bulk signed a contract with the Chinese shipbuilding company Bohai Shipbuilding Heavy Industry for the construction of four 388,000-tonne VLOC. Although initially scheduled for delivery in 2010, the first vessel, *Berge Everest*, was delivered on 23 September 2011, which was followed by the *Berge Aconcagua* on 15 March 2012 and *Berge Jaya* on 12 June 2012. The remaining ship, *Berge Neblina*, was initially also scheduled to be delivered in 2012, but entered service on 4 January 2013. Had Vale not ordered the Valemax fleet in 2008, these ships would have become the largest bulk carriers in the world, surpassing Berge Bulk's own *Berge Stahl*. The four ships have since been chartered by Vale and, despite slight differences in design and contract date predating that of the ships ordered by Vale, they are also referred to as Valemax vessels.

In March 2016, it was reported that three Chinese companies China Ocean Shipping Company (COSCO), China Merchants Energy Shipping and Industrial, and Commercial Bank of China (ICBC) had ordered 10 Valemax vessels each from four Chinese shipyards with a total price of US$2.5 billion. The Valemax ships ordered by China Merchants Energy Shipping would be built by Shanghai Waigaoqiao Shipbuilding (four ships), Qingdao Beihai Shipbuilding (four ships) and China Merchants Group-controlled China Merchants Heavy Industry (Jiangsu) (two ships). COSCO Shipping Corporation awarded the construction of all its ten 400,000-tonne ore carriers to Shanghai Waigaoqiao Shipbuilding. ICBC, which would later hand over the vessels to China Merchants Energy Shipping, announced that six of its ships would be built by the Chinese privately owned shipyard Yangzijian Shipbuilding, while the remaining four would be awarded to Qingdao Beihai Shipbuilding. The first of the 30 Chinese-built second-generation Valemax vessels, *Yuan He Hai*, was delivered on 11 January 2018 and the last one of the series, *Yuan Qian Hai*, in January 2020. In December 2016, the Japanese shipping company NS United ordered a single 400,000 metric tonnes dwt VLOC from Japan Marine United after signing a 25-year contract with Vale. The vessel, which would become the first Valemax ship built in Japan, was delivered in December 2019 as *NSU Carajas*. A second vessel, *NSU Brazil*, was ordered in June 2017 and a third (*NSU Tubarao*) later that year; both were delivered in late 2020. The new ships are larger than the previous record holder, the 364,767 tonne *Berge Stahl*, which had been the largest bulk carrier in the world since it was built in 1986. While the draught of the old vessel is the same as that of the Valemax vessels (23 m, 75 ft), the new ships are 20 m (66 ft) longer and 1.5 m (4.9 ft) wider than the *Berge Stahl* and can carry about 10% more cargo. The Valemax vessels are also the second largest ships in service by dwt tonnage, second only to the TI-class super tankers, which have a dwt tonnage of over 440,000 metric tonnes. Even so, they are still far from the largest ship ever constructed, the *Seawise Giant*, which was built in 1979 and scrapped as the *Knock Nevis* in 2009. The *Seawise Giant* had a length overall of 458.46 m (1504.1 ft) and had a dwt of 564,650 metric tonnes.

Very large and ultra large crude carriers

VLCC have a dwt tonnage or cargo carrying capacity of between 250,000 and 320,000 metric tonnes. ULCC have a dwt of between 320,000 and 550,000 metric tonnes. Worldwide, VLCCs and ULCCs carry between them approximately two billion barrels of oil each year.

SHIP CLASSIFICATION BY TYPE

We have already touched on some of the main types of ships that make up the global merchant fleet such as oil tankers, container ships and bulk carriers. There are dozens of types of ships, and hundreds of sub-types. It would be a mammoth task to describe them all so we will focus instead on the main classifications of ships by type. To start with, we need to differentiate between naval vessels and civilian vessels.

Naval vessels

Naval or military vessels are those which provide the defence of a country's sovereignty and protect the nation's exclusive economic zone (EEZ) and overseas maritime interests. Most coastal states have some form of maritime defence force. In the United Kingdom, for example, this is the Royal Navy. The Royal Navy consists of both military class vessels such as warships and submarines, and civilian manned vessels, which serve as supply ships in the Royal Fleet Auxiliary (RFA). Although civilian crewed, RFA ships are still classified as naval vessels as they are always armed, even if only for self-defence. Because military class vessels do not form part of the global merchant fleet, we will not dwell too much on these types of ships, although we will examine some of the most common forms of warships in later chapters. In terms of merchant vessel classifications, we can categorise ships according to the primary characteristics of their cargo. By this, we mean we must differentiate between wet and dry cargoes, stationary and self-moving cargoes, cargoes that are free flowing and contained, dangerous and non-dangerous cargoes, and vessels that conduct specific roles and functions other than transporting cargoes.

Wet cargoes

Ships that are designed to carry wet cargoes are typically called tankers. These are large, fully sealed vessels consisting of large, compartmentalised tanks. Tankers can carry almost any type of liquid cargo from crude and refined oil products, to liquefied natural and petroleum gas, chemicals, fruit juices, wine and even water. As we saw above, there are many diverse types of tankers based on their size and load carrying capacity.

It is worth mentioning now – although we will cover this again in more detail later – that there are four main types of tankers: oil and product tankers, gas carriers, chemical tankers, and foodstuff tankers. Due to the type of cargo carried, a ship designated as one type of tanker cannot operate as another to prevent cross-contamination.

Dry cargoes

Dry cargo ships are every other type of ship that is designed to carry cargo that is not liquified. This is an overly broad definition and could potentially cover every type of ship from general-purpose, to bulk carriers, to car carriers, to livestock carriers, and everything else in between. Because this is such a broad definition, it is necessary to separate it down into more refined classifications.

Vehicle carriers

Although vehicles are classified as dry cargo, they are in themselves a distinct class of vessel in terms of their size and design characteristics. Vehicle carriers are often classified as either pure car carriers (PCCs) – that is they only carry cars or small commercial vehicles – and pure car and truck carriers (PCTCs). These vessels are capable of transporting cars, commercial vehicles, including trucks and moving stock, and outsized industry equipment such as earth movers and other vast sized mining equipment.

Passenger ships

Although passengers are, in effect cargo, this is inelegant and so to avoid such a clumsy definition, we should differentiate between ships that carry cargo and ships that carry people. The former are cargo ships, and the latter are passenger ships. Passenger ships include a wide variety of different ship types from walk-on only passenger ferries, to roll on roll off ferries, cruise ferries, and cruise ships.

Fishing vessels

As the name suggests, fishing vessels are specially designed ships used for fishing. Fishing vessels may operate within a country's inshore area, in which case they tend to be quite small and return to port the same day; or operate further out to sea returning to port once every couple of days. These craft tend to be called fishing boats. Larger fishing vessels which operate far out at sea for extended periods of time are often classed as trawlers. These ships process many thousands of metric tonnes of fish and operate like floating factories. The largest fishing vessels are actual floating factories, processing tens of thousands of metric tonnes of fish and shellfish every day.

Offshore oilfield vessels

Offshore oilfields need constant supplies of food and equipment to keep operating, often in some of the harshest and most remote areas of the globe. Oil platforms are supplied by offshore support vessels, which sail from their home port to the platform and back again in a continuous cycle. Anchor handling tugs are a type of offshore oilfield vessel, but unlike support or supply vessels, these ships are fitted with extremely sturdy winches and chains designed to pull floating platforms from one location to another. Oilfield operations are complex and there are many types of unique vessels needed to ensure the oilfield remains operational. Drilling ships, diving support vessels, fireboats, and seismic vessels all work together to ensure free flow of oil and gas, powering the global economy.

Construction vessels

Construction vessels are a category of unusual ships, which are specially designed to conduct complex tasks at sea such as dredging, pipelaying, cable laying, and mining.

Harbour support boats

Although not strictly ships, tugboats and pilot boats are essential to the global merchant fleet. Without them, most merchant ships would be unable to navigate into and out of the port. Pilot boats are small inshore boats that ferry harbour pilots from shore to the incoming vessel. The pilot then assists the ship's captain to bring the vessel safely alongside. Many harbours and ports require ships over a certain size to have a pilot on board when entering or leaving the harbour or port limits. Likewise, larger sized vessels often lack maneuverability and require tugboat assistance to turn or berth.

Specialist vessels

Specialist vessels are every other type of vessel not covered elsewhere. This includes icebreakers, crane vessels, heavy lift vessels, nuclear waste carriers and hospital or mercy ships.

SHIP CLASSIFICATION BY GEARS

Some ships are fitted with their own cargo loading and unloading equipment or cranes. These cranes are called gears. A ship that has gears is called a 'geared' ship. Geared ships tend to be smaller in size and are always, albeit not exclusively, bulk carriers. Geared ships are more expensive to buy and maintain as they require specially trained gear operators; however, by having their own gears, geared ships can operate in ports where there are no available loading and unloading facilities. Ships without their own gear are called 'ungeared' ships. Ungeared ships make up most vessels in the merchant fleet.

We have now covered the basic principles of ship classifications. We have seen that there are quite a few ways of classifying ships including whether they are civilian or military, geared or ungeared, carry passengers or cargo, and whether the cargo carried is wet or dry, contained, or free flowing. Ship classification is quite a detailed and complex area, with many other factors determining the true class of the ship, for instance whether the vessel can operate in the polar regions (ice class vessels), can carry more than 12 passengers or has a fuel-type designation (low sulphur fuel oil, heavy sulphur fuel oil, methane powered, etc.). In Part I of this book, we will explore the concept of dry goods vessels and the several types of ships that make up this diverse sector of the global fleet.

Part I

Dry cargo ships

Chapter 1

Bulk carriers

A bulk carrier or bulker is a merchant ship specially designed to transport unpackaged bulk cargo, such as grain, coal, ore, steel coils, and cement, in its cargo holds. Since the first specialised bulk carrier was built in 1852, economic forces have led to continued development of these ships, resulting in increased size and sophistication. Today's bulk carriers are specially designed to maximize capacity, safety, efficiency, and durability. Bulk carriers make up 21% of the world's merchant fleets and range in size from single-hold mini-bulk carriers to mammoth ore ships able to carry 400,000 metric tonnes dwt. Several specialised designs exist. Some can unload their own cargo, some depend on port facilities for unloading, and some even package the cargo as it is loaded. Over half of all bulk carriers have Greek, Japanese, or Chinese owners and more than a quarter are registered in Panama. South Korea is the largest single builder of bulk carriers, and 82% of these ships were built in Asia. On board bulk carriers, the crew are responsible for operation, management, and maintenance of the vessel, taking care of safety, navigation, maintenance, and cargo care, in accordance with international maritime legislation. Cargo loading operations vary in complexity and loading and discharging cargo can take as long as several days. Bulk carriers can be gearless (dependent on terminal equipment) or geared. Crew numbers range in size from three people, on the smallest ships, to over 30, on the largest. Bulk cargo can be very dense, corrosive, or abrasive. This can present safety problems: cargo shifting, spontaneous combustion, and cargo saturation can threaten the ship. The use of ships that are old and have corrosion problems was linked to a spate of bulk carrier sinkings in the 1990s, due to the design of the large hatchways of bulk carriers. These hatchways are important for efficient cargo handling but allow the entry of large volumes of water in heavy seas or stormy weather or if the ship is endangered by sinking. New international regulations have since been introduced to improve ship design and inspection, and to streamline the process for the crew to abandoning ship (Figure 1.1).

DOI: 10.1201/9781003342366-3

Figure 1.1 Bulk carrier *Poseidon*, Kwinana Bulk Jetty, Port of Fremantle, Australia.

The term bulk carrier has been defined in varying ways. As of 1999, the *International Convention for the Safety of Life at Sea* (SOLAS) defined a bulk carrier as 'a ship constructed with a single deck, top side tanks and hopper side tanks in cargo spaces and intended to primarily carry dry cargo in bulk; an ore carrier; or a combination carrier'. Most classification societies use a broader definition, by which a bulk carrier is any ship that carries dry unpackaged goods. Multipurpose cargo ships can not only carry bulk cargo but can also carry other cargoes and are not specifically designed for bulk carriage. The term 'dry bulk carrier' is used to distinguish bulk carriers from bulk liquid carriers such as oil, chemical, or liquefied petroleum gas carriers. Small bulk carriers are almost indistinguishable from general cargo ships, and they are often classified more on the ship's use than its design. Various acronyms are used to describe bulk carriers. 'OBO' describes a bulk carrier that carries a combination of ore, bulk, and oil, and 'O/O' is used for combination oil and ore carriers. The terms 'VLOC', 'VLBC', 'ULOC', and 'ULBC' for very large and ultra-large ore and bulk carriers were adapted from the super tanker designations 'Very Large Crude Carrier' (VLCC) and Ultra Large Crude Carrier (ULCC).

DEVELOPMENT OF THE BULK CARRIER

Before specialised bulk carriers were developed, shippers had two methods to move bulk goods by ship. In the first method, longshoremen loaded the cargo into sacks, stacked the sacks onto pallets, and put the pallets into the cargo hold with a crane. The second method required the shipper to charter an entire ship and spend time and money to build plywood bins into the holds. Then, to guide the cargo through the small hatches, wooden feeders and shifting boards had to be constructed. These methods were slow and extremely labour-intensive. As with the container ship, the problem of efficient loading and unloading has driven the evolution of the bulk carrier. Specialised bulk carriers began to appear as steam-powered ships became more popular. The first steam ship recognised as a bulk carrier was the British collier *John Bowes*, built in 1852. She featured a metal hull, a steam engine, and a ballasting system, which used seawater instead of sandbags. These features helped her succeed in the competitive British coal market. The first self-unloader was the lake freighter *Hennepin* in 1902, which operated on the Great Lakes. This decreased the unloading time of bulk carriers by using a conveyor belt system to move the cargo. The first bulk carriers with diesel propulsion began to appear in 1911. Before World War II, the international shipping demand for bulk products was low – about 25 million metric tonnes for metal ores. Most of this trade was coastal. However, on the Great Lakes, bulk carriers hauled vast amounts of ore from the northern mines to the steel mills on the eastern seaboard. In 1929, 73 million metric tonnes of iron ore were transported via the Lakes, and an almost equal amount of coal, limestone, and other products were also shipped. During this period, two defining characteristics of bulk carriers were beginning to emerge: the double bottom, which was adopted in 1890, and the triangular structure of the ballast tanks, which was introduced in 1905. After World War II, an international bulk trade began to develop among the industrialised nations, particularly between the European countries, the United States, and Japan. Due to the economics of this trade, ocean bulk carriers became larger and more specialised. In this period, Great Lakes freighters increased in size, to maximise the economies of scale, and self-unloaders became more common to cut turnaround time. The thousand-footers of the Great Lakes fleets, built in the 1970s, were among the longest ships afloat, and, in 1979, a record of 214 million metric tonnes of bulk cargo were moved on the Great Lakes.

CATEGORIES

There are different methods used for categorising bulk carriers. The first is to segregate bulk carriers into one of six major size categories: Small, Handysize, Handymax, Panamax, Capesize, and Very Large. Very large bulk and ore carriers fall into the Capesize category but are often considered separately. The second method categorises bulk carriers by regional trade. These are the:

- *Kamsarmax*. Length overall 229 m (751 ft), refers to a new type of ship, larger than the Panamax, that are suitable for berthing at the Port of Kamsar in the Republic of Guinea, where the major loading terminal of bauxite is restricted to vessels not more than 229 m (751 ft).
- *Newcastlemax*. Maximum beam of 50 m (164 ft) and maximum length overall of 300 m (984 ft). This refers to the largest vessel able to enter the Port of Newcastle, Australia, at about 185,000 dwt.
- *Setouchmax*. About 203,000 dwt, being the largest vessels able to navigate the Setouchi Sea, Japan.
- *Seawaymax*. Length overall of 226 m (741.4 ft) and maximum draught of 7.92 m (25.98 ft). Refers to the largest vessels that can pass through the canal locks of the St. Lawrence Seaway (Great Lakes, Canada).
- *Malaccamax*. Length overall 330 m (1,082 ft) and maximum draught of 20 m (65.61 ft) and maximum 300,000 dwt refers to the largest vessels that can pass through the Straits of Malacca.
- *Dunkirkmax*. Length overall of 289 m (948 ft) and maximum allowable beam 45 m (147 ft) equal to a maximum dwt of 175,000 metric tonnes for the eastern harbour lock at the Port of Dunkirk, France.

Mini-bulk carriers are prevalent in the category of small vessels with a capacity of under 10,000 metric tonnes dwt. Mini-bulk carriers carry from 500 to 2,500 metric tonnes, have a single hold, and are designed primarily for river transport. They are often built to pass under bridges and have small crews of between three and eight people. Handysize and Handymax ships are general purpose in nature. These two segments represent 71% of all bulk carriers over 10,000 dwt and have the highest rate of growth. This is partly due to new regulations coming into effect, which put greater constraints on the building of larger vessels. Handymax ships are typically 150–200 m (492 ft–656 ft) in length and 52,000–58,000 metric tonnes dwt, with five cargo holds and four cranes. These ships are usually general purpose in nature. The size of a Panamax vessel is limited by the Panama Canal lock chambers, which can accommodate ships with a beam of up to 32.31 m (106 ft), a length overall of up to 294.13 m (964.99 ft), and a draught of up to 12.04 m (39.5 ft). Capesize ships are too large to traverse the Panama Canal and must round Cape Horn to travel between the Pacific and Atlantic Oceans. Prior to enlargement and dredging of the Suez Canal, Capesize ships could not traverse the Suez and so they needed to go around the Cape of Good Hope. Recent deepening has permitted ships with a draught of up to 20 m (66 ft). Capesize bulk carriers are highly specialised: 93% of their cargo is iron ore and coal. Some ships on the Great Lakes Waterway exceed Panamax dimensions, but they are limited to use on the Great Lakes, as they cannot pass through the smaller St. Lawrence Seaway to the Atlantic Ocean. Very large ore carriers and very large bulk carriers are a subset of the Capesize category reserved for vessels over 200,000 dwt. Carriers of this size are always designed to carry iron ore.

An ore-bulk-oil carrier, also known as combination carrier or OBO, is a ship designed to be capable of carrying wet or dry cargoes. The idea is to reduce the number of empty (ballast) voyages, in which large ships only carry a cargo one way and return empty for another. These are a feature of the larger bulk trades (e.g., crude oil from the Middle East, iron ore and coal from Australia, South Africa, and Brazil). The idea of the OBO was that it would function as a tanker when the tanker markets were good and a bulk/ore carrier when those markets were good. It would also be able to take 'wet' cargo (oil) one way and 'dry' cargo (bulk cargoes or ore) the other way, thus reducing the time it had to sail in ballast (i.e. empty). The first OBO carrier was the *Næss Norseman*, built at A.G. Weser for the company Norness Shipping, owned by the Norwegian shipowner Erling Dekke Næss. Næss and his chief naval architect, Thoralf Magnus Karlsen, were instrumental in conceiving this new type of vessel. *Næss Norseman* was delivered in November 1965 and was 250 m (820 ft) long with a beam of 31.6 m (104 ft), a draught of 13.5 m (44 ft), and a gross register tonnage of 37,965 metric tonnes dwt. OBO carriers quickly became popular among shipowners around the world, and as of 2021, several hundred such vessels have been built albeit the OBO carrier had its glory days in the early 1970s. By the 1980s, it became clear that the type required more maintenance than other vessels, as it was prohibitively expensive to 'switch' from wet to dry cargoes and took valuable time. Moreover, the OBO had much less utility than originally conceived. If the vessel had carried oil, it could switch to carrying ore or other dirty bulk cargoes, but not grain or other clean bulk cargoes. As the 1970s cohort of OBO carriers aged, most of them switched to being used either as pure tankers or as pure ore carriers. By 2021, OBO carriers were no longer as common as they were in the 1970s and 1980s. With few of them being ordered after the 1980s, most existing vessels aged past their design lifetime and no longer exist. Some shipowners continued to support the OBO carrier concept and its trading flexibility. SKS, part of the Kristian Gerhard Jebsen Group, is today operating the largest OBO-fleet in the world consisting of 10 OBO carriers. The latest OBO carrier in the fleet, *D Whale*, was delivered from Hyundai Heavy Industries in 2010. The design of these vessels has been significantly improved compared with the vessels built in the 1970s, and all problems, which were related to the OBO carrier concept – including many that were common amongst tankers at that time – have since been designed out. In the 1990s, a smaller number of OBOs from 70,000 to 100,000 metric tonnes dwt were built for Danish and Norwegian shipowners. A fleet of smaller, 'river-sized' (typically no more than several thousand metric tonnes) ore-bulk-oil carriers were built for use on European Russia's waterways, primarily by the shipowner Volgotanker.

Geared bulk carriers are typically in the Handysize to Handymax size range although there are a small number of geared Panamax class vessels. Like all bulk carriers, they feature a series of holds covered by prominent hatch covers. They have cranes, derricks or conveyors that allows the vessel to load or discharge its cargo in ports without shore-based equipment. This gives geared bulk carriers flexibility in the cargoes they carry and the routes they can travel. Gearless carriers are bulk carriers without cranes or conveyors. These ships depend on shore-based equipment at their ports of call for loading and discharging. They range across all sizes. The larger bulk carriers (VLOCs) can only dock at the largest ports, with some designed with a single port-to-port trade in mind. The use of gearless bulk carriers avoids the costs of installing, operating, and maintaining cranes. Self-dischargers are bulk carriers with conveyor belts, or with the use of an excavator, which is fitted on a traverse running over the vessel's entire hatch, and that can move in a sideways direction as well. This allows them to discharge their cargo quickly and efficiently. Lakers are the bulk carriers prominent on the Great Lakes between North America and Canada. They are often identifiable by having a forward house that helps in transiting the locks. Operating in fresh water, these ships suffer much less corrosion damage and have a much longer lifespan than saltwater ships. As of 2005, there were 98 lakers of 10,000 dwt or over. BIBO or 'Bulk In, Bags Out' bulk carriers are equipped to bag cargo as it is unloaded. The *CHL Innovator*, shown below, is a BIBO bulk carrier. In 1 hour, this ship can unload 300 tonnes of bulk sugar and package it into 50 kg (110 lb) sacks (Figure 1.2).

Figure 1.2 CHL Innovator, leaving the Port of Gdansk, Poland.

FLEET CHARACTERISTICS

The world's bulk transport has reached immense proportions: in 2005, 1.7 billion metric tonnes of coal, iron ore, grain, bauxite, and phosphate were transported by ship. Today, the world's bulk carrier fleet includes 6,225 ships of over 10,000 dwt and represents 40% of all ships in terms of tonnage and 39.4% in terms of vessels. Including smaller ships, bulk carriers have a total combined capacity of almost 346 million dwt. Combined carriers are a small portion of the fleet, representing less than 3% of overall capacity. In 2020, the lake freighters of the Great Lakes, with 98 ships of 3.2 million total dwt metric tonnes, despite forming a small fraction of the total fleet by tonnage and only operating 10 months a year, carried one-tenth of the world's bulk cargo because of the short trip distance and fast turnarounds. As of 2018, the average bulk carrier was just over 13 years old. About 41% of all bulk carriers were less than 10 years old, 33% were over 20 years old, and the remaining 26% were between 10 and 20 years of age. All the 98 bulk carriers registered in the Great Lakes trade are over 20 years old and the oldest still sailing in 2009 was 106 years old.

Flag states

As of 2020, the United States Maritime Administration counted 6225 bulk carriers of 10,000 dwt or greater worldwide. Panama has the highest number of bulk carriers registered with 1,703 ships, more than any four other Flag States combined. In terms of the number of bulk carriers registered, the top five Flag States also include Hong Kong with 492 ships, Malta (435), Cyprus (373), and China (371). Panama also dominates bulk carrier registration in terms of deadweight tonnage. Positions two through five are held by Hong Kong, Greece, Malta, and Cyprus.

Largest fleets

Greece, Japan, and China are the top three owners of bulk carriers, with 1,326, 1,041, and 979 vessels, respectively. These three nations account for over 53% of the world's bulk carrier fleet. Several companies have large private bulk carrier fleets. The multinational company Gearbulk Holding Ltd. has over 70 bulk carriers. Fednav Group in Canada operates a fleet of over 80 bulk carriers, including two designed to work in Arctic ice. Croatia's Atlantska Plovidba d.d. has a fleet of 14 bulk carriers. The H. Vogemann Group in Hamburg, Germany, operates a fleet of 19 bulk carriers. Portline of Portugal owns 10 bulk carriers. Dampskibsselskabet Torm in Denmark and Elcano in Spain also own notable bulk carrier fleets. Other companies specialise in mini-bulk carrier operations: England's Stephenson Clarke Shipping Limited owns a fleet of eight mini-bulk carriers and five small

Handysize bulk carriers, and Cornships Management and Agency Inc., registered in Turkey, owns a fleet of seven mini-bulk carriers.

Ship builders

As with all ship types and categories, Asian companies dominate the construction of bulk carriers. Of the world's 6,225 bulk carriers, almost 62% were built in Japan by shipyards such as Oshima Shipbuilding and Sanoyas Hishino Meisho. South Korea, with notable shipyards owned by Daewoo and Hyundai Heavy Industries, ranked second among builders, with 643 new buildings between them. China, with large shipyards, including Dalian, Chengxi, and Shanghai Waigaoqiao, ranked third, with 509 ships. Taiwan, with shipyards such as China Shipbuilding Corporation, ranked fourth, accounting for 129 new buildings. Shipyards in these top four countries have built over 82% of the bulk carriers afloat in 2022.

Freight charges

Several factors affect the cost to move bulk cargo by ship. The bulk freight market is very volatile. Factors that influence fluctuations include the type of cargo, the ship's size, and the route taken. Moving a Capesize load of coal from South America to Europe can cost anywhere from US$15 to US$25 per tonne. Hauling a Panamax-sized load of aggregate materials from the Gulf of Mexico to Japan can cost anywhere between US$40 per tonne to as much as US$70 per tonne. Some shippers choose instead to charter a ship, paying a daily rate instead of a set price per tonne. In 2005, the average daily rate for a Handymax ship varied between US$18,000–US$30,000. Alternatively, a Panamax ship could be chartered for between US$20,000–US$50,000 per day, and a Capesize for between US$40,000–US$70,000 per day.

Ship breaking

Ships are removed from the fleet by going through a process of scrapping. Shipowners and scrap metal buyers negotiate scrap prices based on factors such as the ship's empty weight (the light tonne displacement or LDT) and prices in the scrap metal market. In 1998, almost 700 ships were scrapped in places like Alang, India, and Chittagong, Bangladesh. This is often done by 'beaching' the ship on open sand, then cutting it apart by hand with gas torches. This is a particularly dangerous operation that often results in serious injuries and fatalities, as well as exposure to toxic materials such as asbestos, lead, and various toxic chemicals. In 2004, some 500,000 metric tonnes of bulk carrier dwt were scrapped, accounting for 4.7% of that year's entire scrapping, with an average price of between US$340 and $350 per LDT tonne.

OPERATIONS

The crew on a bulk carrier typically consists of between 20 and 30 people, though smaller ships can be managed by as few as eight, and the smallest by only three people. The crew includes the captain or master, the deck department, the engine department, and the steward's department. The practice of taking passengers on board cargo ships, once almost universal, is exceedingly rare today and almost nonexistent on bulk carriers. *The International Ship and Port Facility Security* (ISPS) *Code* has made it even more difficult for passengers to travel on board merchant vessels, though some companies still offer travel to senior crew member's families. During the 1990s, bulk carriers contributed to an alarming number of shipwrecks. This led shipowners to commission a study seeking to explain the effect of numerous factors on the crew's effectiveness and competence. The study showed that crew performance on board bulk carriers was the lowest of all groups studied. Among bulk carrier crews, the best performance was found on board younger and larger ships. Crews on better-maintained ships performed better, as did crews on ships where fewer languages were spoken. Fewer deck officers are employed on bulk carriers than on similarly sized ships of other types. A mini-bulk carrier carries on average two to three deck officers, while larger Handysize and Capesize bulk carriers typically carry four. Liquid natural gas tankers of the same size have an additional deck officer and unlicensed mariner.

A bulk carrier's voyages are determined by market forces; routes and cargoes vary significantly. A ship may engage in the grain trade during the harvest season and later move on to carry other cargoes or work on a different route. On board coastal carriers engaged in the tramp trade, the crew will often not know the next port of call until the cargo is fully loaded. Because bulk cargo is so difficult to discharge, bulk carriers spend more time in port than most other ships. A study of mini-bulk carriers found that it takes on average twice as much time to unload a ship as it does to load it. On average, it was found that a mini-bulk carrier spends 55 hours at a time in port, compared with 35 hours for a lumber carrier of comparable size. This time in port increases to 74 hours for Handymax and 120 hours for Panamax class vessels. Compared with the 12-hour turnarounds common for container ships, 15-hour turnarounds for car carriers, and 26-hour turnarounds for large tankers, bulk carrier crews have more opportunities to spend time ashore.

Loading and unloading a bulk carrier is time-consuming and dangerous. The process is planned by the ship's chief officer (chief mate) under the direct and continued supervision of ship's captain (or master). International regulations require that the captain and terminal master agree on a detailed plan before loading and unloading operations begin. Deck officers and stevedores then oversee the operations. Occasionally, it has been known for loading errors to cause a ship to capsize or break in half at the quayside. The loading method used depends on both the cargo and the equipment available on the ship and on the dock. In the least advanced ports, cargo is often loaded with shovels or bags poured from the hatch cover, although this system is being replaced with faster, less labour-intensive methods. Double-articulation cranes, which can load at a rate of 1,000 metric tonnes per hour, represent the most widely used method, and the use of shore-based gantry cranes, reaching 2,000 metric tonnes per hour, is growing. A crane's discharge rate is limited by the bucket's capacity (from 6 to 40 metric tonnes) and by the speed at which the crane can take a load, deposit it at the terminal and return to take the next load. For modern gantry cranes, the total time of the grab-deposit-return cycle is about 50 seconds. Conveyor belts offer a very efficient method of loading, with standard loading rates varying between 100 and 700 metric tonnes per hour, although the most advanced ports can offer rates of 16,000 metric tonnes per hour. Start-up and shutdown procedures with conveyor belts, though, are complicated and require time to operate safely. Self-discharging ships use conveyor belts with load rates of around 1000 metric tonnes per hour. Once the cargo is discharged, the crew begins to clean the holds. This is particularly important if the next cargo is of a different type. The immense size of cargo holds and the tendency of cargoes to be physically irritating add to the difficulty of cleaning the holds. When the holds are clean, the process of loading can begin. It is crucial to maintain the cargo level during loading to keep an even keel. As the hold is filled, machines such as excavators and bulldozers are often used to keep the cargo level. Levelling is particularly important when the hold is only partly full since cargo is more likely to shift. Extra precautions are taken, such as adding longitudinal divisions and securing wood atop the cargo. If a hold is full, a technique called 'tomming' is used, which involves digging out a 2 m (6 ft) hole below the hatch cover and filling it with bagged cargo or weights. This helps to lower the ship's centre of gravity and improves stability.

DESIGN AND ARCHITECTURE

A bulk carrier's design is defined by the cargo it carries. The cargo's density, also known as its stowage factor, is the key influence. Densities for common bulk cargoes vary from 0.6 metric tonnes per cubic metre for light grains to three metric tonnes per cubic metre for iron ore. The overall cargo weight is the limiting factor in the design of an ore carrier since the cargo is so dense. Coal carriers, on the contrary, are limited by overall volume since most bulk carriers can be filled with coal before reaching their maximum draught. For a given tonnage, the second factor, which governs the ship's dimensions, is the size of the ports and the waterways they operate in. For example, a vessel that will pass the Panama Canal will be limited in its beam and draught. For most designs, the ratio of length-to-width ranges between 5 m (16.4 ft) and 7 m (22.96 ft), with an average of 6.2 m (20.34 ft). The ratio of length-to-height is between 11 m (36 ft) and 12 m (39 ft).

Machinery

The engine room on a bulk carrier is usually near the stern, under the superstructure. Larger bulk carriers, from Handymax up, typically have a single two-stroke low-speed crosshead diesel engine directly coupled to a fixed-pitch propeller. Electricity is produced by auxiliary generators and/or an alternator coupled to the propeller shaft. On smaller bulk carriers, one or two four-stroke diesels are used to turn either a fixed or controllable-pitch propeller via a reduction gearbox, which may also incorporate an output for an alternator. The average design ship speed for bulk carriers of Handysize and above is 13.5–15 knots (15.5–17.3 mph or 25.0–27.8 km/h). The propeller speed is low, at about 90 revolutions per minute, although this depends on the size of the propeller. As a result of the 1973 oil crisis, the 1979 energy crisis and the resulting rise in oil prices since 2009, experimental designs using coal to fuel ships have been assessed since the late 1970s and early 1980s. Australian National Lines (ANL) constructed two 74,700 ton coal-burner ships called *River Boyne* and *River Embely*. A further two coal-fired bulk carriers were constructed by TNT named *TNT Capricornia* and *TNT Capentaria*, later renamed *Fitzroy River* and *Endeavor River*. These ships were financially effective for the duration of their lives, and their steam engines were able to generate a shaft-power of 19,000 horsepower (14,000 kW). This strategy gave an interesting advantage to carriers of bauxite and similar fuel cargoes but suffered from poor engine yield compared with the higher maintenance cost and efficient modern diesels, maintenance problems due to the supply of ungraded coal and high initial costs.

Hatches

A hatch or hatchway is the opening at the top of the cargo hold. The mechanical devices, which allow hatches to be opened and closed, are called hatch covers. In general, hatch covers form between 45% and 60% of the ship's breadth, or beam, and 57–67% of the length of the holds. To efficiently load and unload cargo, hatches must be large, but large hatches present structural problems. Hull stress is concentrated around the edges of the hatches, which means these areas must be reinforced. Often, hatch areas are reinforced by locally increasing the scantlings or by adding structural members called stiffeners. Both options have the undesired effect of adding extra weight to the ship. As recently as the 1950s, hatches had wooden covers that would be broken apart and rebuilt by hand, rather than opened and closed. Newer vessels have hydraulic-operated metal hatch covers that can often be operated by one person. Hatch covers can slide forwards, backwards, or to the side, lift or fold up. It is essential that the hatch covers are fully watertight; unsealed hatches can lead to accidental cargo hold flooding, which is a major cause for bulk carrier sinkings. Regulations regarding hatch covers have evolved since the investigation into the loss of the British bulk carrier *MV Derbyshire* in 1980. The Load Line Conference of 1966 imposed a requirement that hatch covers be able to withstand a load of 1.74 metric tonnes/m^2 of sea water, and a minimum scantling of 6 mm (0.23 in) for the tops of the hatch covers. This was later increased by the International Association of Classification Societies in their Unified Requirement S21 in 1998. This standard requires that the pressure due to sea water be calculated as a function of freeboard and speed, especially for hatch covers located towards the forward portion of the ship.

Hull

Bulk carriers are designed to be easy to build and to store cargo efficiently. To facilitate construction, bulk carriers are built with a single hull curvature. Also, whilst a bulbous bow allows a ship to move more efficiently through water, bulk carrier designers have moved towards simple vertical bows on larger ships. Full hulls, with large block coefficients, are almost universal, and as a result, bulk carriers are inherently slow. This is offset by their efficiency. Comparing a ship's carrying capacity in terms of deadweight tonnage to its weight when empty is one way to measure a ship's efficiency. A small Handymax ship, for example, can carry as much as five times its weight. In larger designs, this efficiency is even more pronounced. Capesize vessels can carry more than eight times their dwt tonnage. Bulk carriers have a cross-section typical of most merchant ships. The upper and lower corners of the hold are used as ballast tanks, as is the double bottom area. The corner tanks are reinforced and serve another purpose besides controlling the ship's trim. Designers choose the angle of the corner tanks to be less than that of the angle of repose of the anticipated cargoes. This reduces side-to-side movement, or 'shifting', of cargo, which can endanger the ship. The double bottoms are also subject to design constraints. The primary concern is that they are high enough to allow the passage of pipes and cables. These areas must also be sufficiently large enough to allow the crew and contractors safe access to perform surveys and maintenance. Conversely, concerns around excess weight and wasted volume keep the double bottom very tight. Bulk carrier hulls are made from mild steel. Some manufacturers have preferred high-tensile steel recently to reduce the tare weight. However, the use of high-tensile steel for longitudinal and transverse reinforcements can reduce the hull's rigidity and resistance to corrosion. Forged steel is used for some ship parts, such as the propeller shaft support. Transverse partitions are made of corrugated iron, reinforced at the bottom and at connections. More recently, ship designers have investigated the possibility of constructing bulk carrier hulls using a concrete-steel sandwich.

In the late 1990s and early 2000s, double hulls became increasingly popular. Designing a vessel with double sides adds primarily to its breadth, since bulk carriers are already required to have double bottoms. One of the advantages of the double hull is to make room to place all the structural elements in the sides, removing them from the holds. This increases the volume of the holds, and simplifies their structure, which helps in loading, unloading, and cleaning. Double sides also improve the ship's capacity for ballasting, which is useful when carrying light goods. The ship may have to increase its draught for stability or seakeeping reasons, which is done by adding ballast water. A recent design, called Hy-Con, seeks to combine the strengths of single-hull and double-hull construction. Short for Hybrid Configuration, this design doubles the forward-most and rear-most holds and leaves the others single-hulled. This approach increases the ship's solidity at key points, whilst reducing the overall tare weight. Since the adoption of the double hull has been more of an economic than a purely architectural decision, some argue that double-sided ships receive fewer comprehensive surveys and suffer more from hidden corrosion. Despite this opposition, double hulls became a requirement for Panamax and Capesize vessels in 2005. Freighters are in continual danger of 'breaking their backs', and thus, longitudinal strength is a primary architectural concern. A naval architect uses the correlation between longitudinal strength and a set of hull thicknesses called scantlings to manage problems of longitudinal strength and stresses. A ship's hull is composed of individual parts called members. The set of dimensions of these members is called the ship's scantlings. Naval architects calculate the stresses a ship can be expected to be subjected to, add in safety factors, and then can calculate the required scantlings. These analyses are conducted when traveling empty, when loading and unloading, when partially and fully loaded, and under conditions of temporary overloading. Places subject to the largest stresses are studied carefully, such as the hold-bottoms, hatch-covers, bulkheads between holds, and the bottoms of ballast tanks. Great Lakes bulk carriers also must be designed to withstand springing, or developing resonance with the waves, which can cause fatigue fractures. Since 1 April 2006, the International Association of Classification Societies has adopted the Common Structural Rules. The rules apply to bulk carriers more than 90 m (295 ft) in length and require that scantlings' calculations consider factors such as the effect of corrosion, the harsh conditions often found in the North Atlantic, and dynamic stresses during loading. The rules also establish margins for corrosion, from 0.5 to 0.9 mm (0.01–0.03 in).

Safety

The 1980s and 1990s were a very unsafe period for bulk carriers. Many bulk carriers sank during this time: 99 were lost between 1990 and 1997 alone. Most of these sinkings were sudden and quick, making it impossible for the crew to escape. In fact, more than 650 seafarers were lost during this period. Due partly to the sinking of the *MV Derbyshire*, a series of international safety resolutions regarding bulk carriers were adopted throughout the 1990s and 2000s.

Stability problems

Cargo shifting poses a fundamental danger for bulk carriers. The problem is even more pronounced with grain cargoes since grain settles during a voyage and creates extra space between the top of the cargo and the top of the hold. The cargo is then free to move from one side of the ship to the other as the ship rolls. This can cause the ship to list, which, in turn, causes more cargo to shift. This kind of chain reaction can capsize a bulk carrier very quickly indeed. The 1974 SOLAS Convention sought to control this sort of problem. The regulations required the upper ballast tanks to be designed in a manner which could prevent shifting. They also required cargoes to be levelled, or trimmed, using excavators in the holds. The practice of trimming reduces the amount of the cargo's surface area in contact with air. This has a useful side effect, which is reducing the chances of spontaneous combustion in cargoes such as coal, iron, and metal shavings. Another risk that can affect dry cargoes is absorption of ambient moisture. When exceptionally fine concretes and aggregates mix with water, the mud created at the bottom of the hold shifts easily, producing free surface effect. The only way to control these risks is by good ventilation practices and by careful monitoring of the holds for the presence of water. The *International Maritime Solid Bulk Cargoes (IMSBC) Code* was introduced to facilitate the safe stowage and shipment of solid bulk cargoes by providing information on the dangers associated with certain cargoes and instructions on the procedures to be adopted when loading such cargoes.

Structural problems

In the year 1990, 20 bulk carriers were sunk, taking with them 94 crewmembers. In 1991, 24 bulk carriers sank, with the loss of 154 souls. This level of loss focused attention on the safety aspects of bulk carriers. The American Bureau of Shipping concluded that the losses were 'directly traceable to failure of the cargo hold structure' and Lloyd's Register of Shipping added that the hull sides could not withstand 'the combination of local corrosion, fatigue cracking and operational damage'. As more studies were conducted, it became increasingly apparent that certain trends were present in each incident. Investigations sought to demonstrate that sea water was entering the forward hatch because of large waves, a poor seal, or metal and structural corrosion. The extra water weight in hold number one then compromised the partition to hold number two. This led to water entering hold number two, altering the trim of the vessel to such an extent that more water entered the holds. With two holds rapidly filling with water, the bow would submerge, and the ship quickly sinks, leaving little time for the crew to react. Previous practices had required ships to withstand the flooding of a single forward hold but did not guard against situations where two holds would flood. The case where two after (rear) holds are flooded is no better, as this means that the engine room is quickly flooded, leaving the ship powerless and without propulsion. If two holds in the middle of the ship are flooded, the stress on the hull can become so great that the ship snaps in two. In addition to this, various other contributing factors were also identified. Most of the sinkings involved ships that were over 20 years of age. A glut of ships of this age occurred in the 1980s, caused by an overestimate of the growth of international trade. Rather than replacing them prematurely, shipping companies were compelled, on cost grounds, to keep their ageing vessels in service. This, coupled with corrosion due to a lack of maintenance, affected the seals of the hatch covers and the strength of the bulkheads which separated the holds. This corrosion is difficult to detect due to the immense size of the surfaces involved. Advanced methods of loading were not foreseen when the ships were designed. While the new processes are more efficient, loading is more difficult to control (it can take over an hour just to halt the operation), occasionally resulting in overloading the ship. These unexpected shocks, over time, can damage the hull's structural integrity. Recent use of high-tensile steel allows building a structure with less material and weight whilst retaining a similar level of strength. However, because it is thinner than regular steel, high-tensile steel can corrode more easily, as well as it can develop metal fatigue in heavy seas. Considering the number of bulk carriers sinking, and the evidence arising from the incident investigations, Lloyd's Register concluded that the principal cause of these sinkings was the attitude of shipowners, who allowed ships with known problems to sea. New rules were adopted in the 1997 amendments to the SOLAS Convention, which focused on problems such as reinforcing the bulkheads and the longitudinal frame, more stringent inspections (with a particular focus on corrosion) and routine in-port inspections. Moreover, the 1997 additions also required bulk carriers with restrictions (for instance, forbidden from carrying certain types of cargoes) to mark their hulls with large, easy-to-see triangles.

Crew safety

Since December 2004, Panamax and Capesize bulk carriers have been required to carry free-fall lifeboats located on the stern, behind the accommodation block. This arrangement allows the crew to abandon ship quickly in case of an emergency. Free-fall lifeboats have attracted significant criticism. One argument against the use of free-fall lifeboats is that the evacuees require 'some degree of physical mobility, even fitness' to enter and launch the boat. Injuries have occurred during launches, for example, in the event of incorrectly secured safety belts. In December 2002, chapter XII of the

SOLAS Convention was amended to require the installation of high-level water alarms and monitoring systems on all bulk carriers. This safety measure quickly alerts watch standers on the bridge and in the engine room in the event of flooding in the holds. In cases of extreme flooding, it is anticipated that these detectors should quicken the process of abandoning ship.

In this chapter, we have examined ships that are specially designed and constructed to transport vast quantities of free flowing and loose cargo in bulk. These bulk carriers are some of the largest vessels ever to be built, and rival only oil tankers in their length and breadth. In Chapter 2, we will turn our attention to one of the most common types of ships, the container or box ship.

Chapter 2

Container ships

The container ship (also called a box ship or containership) is a cargo ship that carries all its load in truck-size intermodal containers, using a technique called *containerisation*. Container ships are a common means of commercial intermodal freight transport and now carry most seagoing non-bulk cargo. Container ship capacity is measured in 20-foot equivalent units (TEU). Typical loads are a mix of 20-foot (one TEU) and 40-foot (one FEU – broadly the equivalent of two TEU) ISO-standard containers, with the latter being predominant. At present, about 90% of non-bulk cargo worldwide is transported by container ships, and the largest modern container ships can carry up to 24,000 TEU (e.g. the *Ever Ace*). Container ships now rival crude oil tankers and bulk carriers as the largest commercial seaborne vessels in operation. There are two main types of dry cargo, namely bulk cargo and break-bulk cargo. On the one hand, bulk cargoes, such as grain or coal, are transported unpackaged in the hull of the ship, in a large volume. Break-bulk cargoes, on the other hand, are transported in packages, and are manufactured goods. Before the advent of containerisation in the 1950s, break-bulk items were loaded, lashed, unlashed and unloaded from ships one piece at a time. However, by grouping cargo into containers, 28–85 m^3 (1000–3000 cu/ft) of cargo, or up to about 29,000 kg (64,000 lbs) can be moved simultaneously and each container is secured to the ship once in a standardised manner. Containerisation has increased the efficiency of moving traditional break-bulk cargoes significantly, reducing shipping time by as much as 84% and costs by 35%. In 2001, more than 90% of world trade in non-bulk goods was transported in ISO containers. In 2009, almost one quarter of the world's dry cargo was shipped by container, equal to an estimated 125 million TEUs or 1.19 billion tonnes of cargo (Figure 2.1).

The first ships designed to carry standardised load units were used in the late 18th century in England. In 1766, James Brindley designed the box boat *Starvationer* with 10 wooden containers, to transport coal from Worsley Delph in Lancashire to Manchester via the Bridgewater Canal. Before the Second World War, the first container ships were used to carry the baggage of the luxury passenger train *Golden Arrow*/*La Flèche d'Or* from London,

DOI: 10.1201/9781003342366-4

Figure 2.1 Container handling.

England to Paris, France. These containers were loaded in London or Paris and carried to the ports of Dover or Calais on flat railway cars. In February 1931, the world's first container ship was launched; the *Autocarrier*. Owned by the English railway company Southern Railways, the *Autocarrier* had 21 slots for containers. The first container ships to be used after the Second World War were converted from re-purposed surplus war-era T2 tankers. In 1951, the first purpose-built container vessels began operating from Denmark, and between Seattle and Alaska. The first commercially successful container ship was the *Ideal X* (Figure 2.2), a re-purposed T2 tanker, owned by the American transport magnate Malcom McLean. The *Ideal X* was fitted with deck mountings that could secure 58 metal containers. The ship began its maiden voyage as a container ship between Newark, New Jersey and Houston, Texas. In 1955, McLean purchased the Pan Atlantic Steamship Company from Waterman Steamship and adapted its ships to carry cargo in large uniform sized metal containers. By 1956, McLean had built his company, McLean Trucking, into one of the largest freighter fleets in the United States. It was not until May 1964, however, that the world's first fully cellular (FC) purpose-built container ship, the *MV Kooringa*, was launched by the Australian company, Associated Steamships Pty. Ltd., in partnership with McIlwraith, McEacharn & Co.

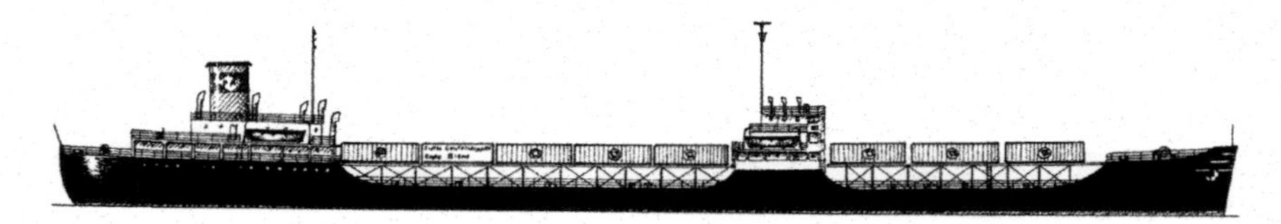

Figure 2.2 Ideal X.

Container ships work by eliminating the individual hatches, holds and dividers of the traditional general cargo vessel. The hull of a typical container ship is a huge warehouse divided into cells by vertical guide rails. These cells are designed to hold cargo in pre-packed units – the containers. Shipping containers are usually made of steel, but other materials like aluminium, fibreglass or plywood may also be used. They are designed to be entirely transferrable from and to smaller coastal carriers, trains, trucks, or semi-trailers. This gives the container its name the *intermodal container*. There are several types of containers, categorised according to their size and functions. Today, about 90% of non-bulk cargo worldwide is transported by container, and modern container ships can carry over 24,000 TEU. As a class, container ships now rival crude oil tankers and bulk carriers as the largest commercial vessels in operation. Although containerisation caused a revolution in the world of shipping, its introduction was not straightforward. The ports and railway companies and shipowners, in general, were concerned about the huge costs of developing the new port and railway infrastructure needed to handle container ships, and for the movement of containers on land by rail and road. Moreover, trade unions were concerned about massive job losses amongst port and dock workers, as containers would eliminate the manual jobs associated with cargo handling. In the event, it took 10 years of legal battles before container ships would be pressed into international service. In 1966, the first container liner service from the United States to Rotterdam commenced.

Importantly, containerisation changed not only the face of shipping, but also revolutionised world trade. A container ship can be loaded and unloaded in a few hours compared with the days it takes for a traditional cargo vessel. This, besides cutting labour costs, has reduced shipping times between ports; for example, it takes a few weeks instead of months for a consignment to be delivered from China to the UK and vice versa. This has also resulted in reduced breakage due to less handling; and less danger of cargo shifting during passage. But the greatest enhancement is, as containers are sealed and only opened at the destination, theft has been reduced. Containerisation has lowered shipping costs and decreased shipping time, and this has in turn helped the growth of international trade. Cargo that once arrived in cartons, crates, bales, barrels, or bags now arrive at the destination factory in sealed containers, with no indication to the human eye of their contents, except for a product code that machines can scan and computers trace. This system of tracking has been so exact that a 2-week voyage can be timed for arrival with an accuracy of under 15 minutes. It has resulted in trade revolutions such as on time guaranteed delivery and just in time manufacturing. Raw materials arrive from factories in sealed containers less than an hour before they are required for manufacture, resulting in reduced inventory costs. Exporters load merchandise in boxes that are provided by the shipping companies. They are then delivered to the port by road, rail or a combination of both for loading onto awaiting container ships. Prior to containerisation, huge gangs of men would spend hours fitting myriad items of cargo into different holds. Today, massive gantry cranes are used to place containers in specific locations on board the ship. When the hull has been fully loaded, additional containers are stacked on the main deck. The largest container ships in operation today measure over 400 m (1300 ft) in length and can carry loads equal to the cargo-carrying capacity of 16–17 pre-World War II freighter ships (Table 2.1).

Table 2.1 Container ship size categories

Class	*Capacity (TEUs)*	*Length*	*Beam*	*Draught*	*Notes*
Ultra Large Container Vessel	14,501–higher	366 m (1,200 ft) and longer	49 m (160.7 ft) and wider	15.2 m (49.9 ft) and deeper	With a length of 400 m (1312 ft), a width of 59 m (193.5 ft), draught of 14.5 m (47.57 ft), and a capacity of 18,270 TEU, ships of the *Mærsk Triple E* class can transit the Suez Canal.
New Panamax or Neo-Panamax	10,000–14,500	366 m (1200 ft)	49 m (160.7 ft)	15.2 m (49.9 ft)	With a beam of 43 m (141.07 ft), ships of the *COSCO Guangzhou class* are too big to fit through the Panama Canal's old locks but could easily fit through the new expansion.
Post-Panamax	5101–10,000				

(*Continued*)

Table 2.1 (*Continued*) Container ship size categories

Class	*Capacity (TEUs)*	*Length*	*Beam*	*Draught*	*Notes*
Panamax	3,001–5,100	294.13 m (965 ft)	32.31 m (106 ft)	12.04 m (39.5 ft)	Ships of the *Bay class* are at the upper limit of the *Panamax class*, with an overall length of 292.15 m (958 ft), beam of 32.2 m (105.64 ft), and maximum depth of 13.3 m (43.63 ft).
Feedermax	2,001–3,000				Container ships under 3,000 TEU are typically called feeders. In some areas of the world, they might be outfitted with cargo cranes.
Feeder	1,001–2,000				
Small Feeder	Up to 1,000				

SIZE CATEGORIES

Container ships are distinguished into seven major size categories: small feeder, feeder, Feedermax, Panamax, Post-Panamax, New Panamax and Ultra Large. As of December 2012, there were 161 container ships in the VLCS class (Very Large Container Ships, i.e. more than 10,000 TEU), and 51 ports in the world, which can accommodate them. The size of a Panamax vessel is limited by the original Panama Canal's lock chambers, which can accommodate ships with a beam of up to 32.31 m (106 ft), a length overall of up to 294.13 m (964.99 ft), and a draught of up to 12.04 m (39.50 ft). The Post-Panamax category has historically been used to describe ships with a moulded breadth over 32.31 m (106.00 ft); however, the Panama Canal expansion project has caused some changes in terminology. The New Panamax category is based on the maximum vessel-size that can transit the latest set of locks, which opened in June 2016. The third set of locks were built to accommodate container ships with a length overall of 366 m (1201 ft), a maximum width of 49 m (161 ft), and a tropical fresh-water draught of 15.2 m (50 ft). New Panamax class vessels are wide enough to carry 19 columns of containers, having a total capacity of approximately 12,000 TEU and of comparable size to a Capesize bulk carrier or a Suezmax tanker. Container ships under 3000 TEU are called feeder ships or feeders.

They are small ships that typically operate between smaller container ports. Some feeders collect their cargo from small ports, drop it off at large ports for trans-shipment on larger ships, and distribute containers from the large port to smaller regional ports. Feeder vessels are the most likely to carry their own deck mounted cargo cranes on board. Lift-on/Lift-off (LOLO, LO/LO or Lo/Lo) ships are cargo ships with on board cranes to load and unload cargo. Ships with cranes or other cargo handling equipment are categorised as geared vessels. As container ships usually have no on-board cranes or other mechanism to load or unload their cargo, they are therefore dependent on dockside container cranes to load and unload. However, lift-on lift-off vessels can load and unload their own cargo unassisted. Lift-on lift-off vessels can operate out of docks with no dockside cargo handling equipment.

CONTAINER SHIP ARCHITECTURE

There are several key points in the design of modern container ships. The hull, like that of bulk carriers and general cargo ships, is built around a strong keel. Into this frame is set one or more below-deck cargo holds, numerous tanks, and the engine room. The holds are topped by hatch covers, onto which more containers may be stacked. Many container ships have cargo cranes installed on them, and some have specialised systems for securing the containers on board. The hull of a modern cargo ship is a complex arrangement of steel plates and strengthening beams. Resembling ribs, and fastened at right angles to the keel, are the ship's frames. The ship's main deck, the metal platework that covers the top of the hull framework, is supported by beams that are attached to the tops of the frames and run the full breadth of the ship. The beams not only support the deck, but along with the deck, frames, and transverse bulkheads, strengthen and reinforce the shell. Another feature of recent hulls is a set of double-bottom tanks, which provide a second watertight shell that runs most of the length of a ship. The double-bottoms hold liquids such as fuel oil, ballast water or fresh water. The ship's engine room accommodates the main engines and auxiliary machinery such as the fresh water and sewage systems, electrical generators, fire pumps, and air conditioners. In most new ships, the engine room is in the aft portion of the hull.

Cargo cranes

A major characteristic of a container ship is whether it has cranes installed for managing its cargo. Those that have cargo cranes are called geared vessels and those that do not are called ungeared or gearless vessels. The earliest purpose-built container ships in the 1970s were all gearless. Since then, the percentage of geared newbuilds has fluctuated widely, but has decreased overall, with only 7.5% of container ship capacity in 2009 being equipped with cranes. While geared container ships are more flexible in that they can visit ports that are not equipped with quayside container cranes, they suffer from several drawbacks. Firstly, geared ships are more expensive to purchase than gearless ships. Secondly, geared ships incur greater recurring expenses, such as maintenance and fuel costs. The *United Nations Council on Trade and Development* (UNCTAD) characterises geared ships as a 'niche market only appropriate for those ports where low cargo volumes do not justify investment in port cranes or where the public sector does not have the financial resources for such investment'. Instead of the typical rotary cranes, some geared ships have gantry cranes installed. These cranes, specialised for container work, can roll forward and aft on rails. In addition to the additional capital expense and maintenance costs, these cranes load and discharge containers much slower than their shoreside counterparts. The introduction and improvement of shoreside container cranes have been a key to the success of the container ship. The first crane to be specifically designed for container work was built at California's Port of Alameda in 1959. By the 1980s, quayside gantry cranes were capable of moving containers on a 3-minute cycle, or up to 400 tonnes per hour. By the mid-2000s, this had improved even more. In March 2010, at the Port Klang in Malaysia, a new world record was set when 734 container moves were made in a single hour. This record was achieved using as many as nine cranes to simultaneously load and unload the *MV CSCL Pusan*, a ship with a capacity of 9600 TEUs. As we said above, feeder ships in the 1500–2499 TEU range are more likely to be fitted with cranes, with more than 60% of this category being geared ships. Slightly less than one-third of the very smallest ships (from 100 to 499 TEU) are geared, and almost no ships with a capacity of over 4000 TEU are geared (Table 2.2).

Table 2.2 Container loading records (2018–2021)

Departure	*Vessel*	*TEU*	*Port*	*Destination*
14 Aug 2021	*Ever Ace*	21,710	Yantian, China	Rotterdam, Netherlands
08 Apr 2021	*CMA CGM Jacques Saadé*	21,433	Singapore	Le Havre, France
12 Oct 2020	*CMA CGM Jacques Saadé*	20,723	Singapore	Le Havre, France
08 May 2020	*HMM Algeciras*	19,621	Yantian, China	Rotterdam, Netherlands
28 Jul 2019	*MSC Gülsün*	19,574	Tanjung Pelepas, Malaysia	Algeciras, Spain
01 Jun 2019	*Monaco Mærsk*	19,284	Tanjung Pelepas, Malaysia	Rotterdam, Netherlands
11 Feb 2019	*MOL Tribute*	19,190	Singapore	Southampton, England
Aug 2018	*Mumbai Mærsk*	19,038	Tanjung Pelepas, Malaysia	Rotterdam, Netherlands

Cargo holds

Efficiency has always been key in the design of container ships. While containers may be transported on conventional break-bulk ships, cargo holds for dedicated container ships are specially constructed to speed loading and unloading, and to efficiently keep containers secure while at sea. A key aspect of container ship specialisation is the design of the hatches, the openings from the main deck to the cargo holds. The hatch openings stretch the entire breadth of the cargo holds and are surrounded by a raised steel structure called the hatch coaming. On top of the hatch coamings are the hatch covers. Until the 1950s, hatches were typically secured with wooden boards and tarpaulins held down with battens. At present, hatch covers may be solid metal plates that are lifted on and off the ship by cranes, or else articulated with mechanisms that are opened and closed using powerful hydraulic rams. Another key component of dedicated container-ship design is the use of cell guides. Cell guides are strong vertical structures constructed of metal, which are installed into the ship's cargo holds. These structures guide containers into well-defined rows during loading and provide some support for containers against the ship's rolling in heavy seas. So fundamental to container ship design are cell guides that UNCTAD uses their presence to distinguish dedicated container ships from general break-bulk cargo ships. A system of three dimensions is used in cargo plans to describe the position of a container aboard the ship. The first coordinate is the bay, which starts at the front of the ship and increases aft. The second coordinate is the row. Rows on the starboard side are given odd numbers and those on the port side are given even numbers. The rows nearest the centreline are given small numbers, and the numbers increase for slots further from the centreline. The third coordinate is the tier, with the first tier at the bottom of the cargo holds, the second tier on top of that, and so forth upwards. Container ships only take 20-foot, 40-foot, and 45-foot containers. Forty-five footers can only fit above deck. 40-foot containers are the primary container size, making up about 90% of all container shipping. Since container shipping moves 90% of the world's freight, over 80% of the world's freight moves via 40-foot containers (Figure 2.3).

Lashing systems

Numerous systems are used to secure containers aboard ships, depending on factors such as the type of ship, the type of container, and the location of the container. Stowage inside the holds of FC ships is the most straightforward. These ships typically use simple metal forms called container guides, locating cones, and anti-rack spacers to lock the containers together. Above-decks, without the extra support of the cell guides, more complicated equipment must be used. Three types of systems are currently in wide use: lashing systems, locking systems, and buttress systems. Lashing systems secure containers to the ship using devices made from wire rope, rigid rods, or

Figure 2.3 Inside the cargo hold on a container ship.

chains and devices to tension the lashings, such as turnbuckles. The effectiveness of lashings is increased by securing the containers to each other, either by simple metal forms (such as stacking cones) or more complicated devices such as twist-lock stackers. A typical twist-lock is inserted into the casting hole of one container and rotated to hold it in place, and then another container is lowered on top of it. The two containers are locked together by twisting the handle of the device. A typical twist-lock is constructed of forged steel and ductile iron and has a shear strength of forty-eight tonnes. The buttress system, used on larger container ships, uses a system of large towers attached to the ship at both ends of each cargo hold. As the ship is loaded, a rigid, removable stacking frame is added, structurally securing each tier of containers together (Figure 2.4).

The Bridge

Container ships have typically had a single bridge and accommodation unit towards the stern, but to reconcile demand for larger container capacity with SOLAS visibility requirements, several contemporary designs have been developed. As of 2015, most large container ships have the bridge positioned further forward, usually just behind the midships, and separate from the exhaust stack. Some smaller container ships working in European ports and rivers have liftable wheelhouses, which can be lowered to pass under low bridges.

Figure 2.4 Typical container lashing system on *MV Pollux* at Torshaven, Iceland.

CONTAINER FLEET CHARACTERISTICS

As of 2010, container ships made up 13.3% of the world's fleet in terms of dwt tonnage. The global total of container ship dwt tonnage has increased from 11 million dwt in 1980 to 169.0 million dwt in 2010. The combined dwt tonnage of container ships and general cargo ships, which also often carry containers, represents 21.8% of the world's merchant fleet. As of 2009, the average age of container ships worldwide was 10.6 years, making them the youngest general vessel type, followed by bulk carriers at 16.6 years, oil tankers at 17 years, general cargo ships at 24.6 years, and others at 25.3 years. Most of the global carrying capacity in FC container ships is in the liner service, wherein ships trade on scheduled routes. As of January 2020, the top 20 liner companies controlled 67.5% of the world's FC container capacity, with 2673 vessels of an average capacity of 3,774 TEU. The remaining 6,862 FC ships have an average capacity of 709 TEU each. Most of the capacity of FC container ships used in the liner trade are owned by German shipowners, with approximately 75% alone owned by Hamburg brokers. It is a customary practice for the large container lines to supplement their own ships with chartered-in ships, for example in 2009, 48.9% of the tonnage of the top 20 liner companies was chartered-in in this manner (Table 2.3).

Table 2.3 Largest container ship operators (2021)

Rank	*Company*	*Country of registration*	*Fleet capacity*
1	Mærsk Line	Denmark	4,121,789
2	MSC	Switzerland / Italy	3,920,784
3	CMA CGM	France	3,049,743
4	COSCO	China	3,007,421
5	Hapag Lloyd	Germany	1,789,399
6	ONE	Japan	1,600,531
7	Evergreen	Taiwan	1,345,537
8	Hyundai Merchant Marine	South Korea	752,604
9	Yang Ming	Taiwan	628,463
10	ZIM Integrated Shipping	Israel	409,810

Flag states

International maritime law requires that every merchant ship be registered in a country, called its Flag State. A ship's Flag State exercises regulatory control over the vessel and is required to inspect it regularly, certify the ship's equipment and crew, and issue safety and pollution prevention documents. As of 2006, the United States Bureau of Transportation Statistics counted 2837 container ships of 10,000 long tonnes dwt or greater in global operation. Panama is the world's largest Flag State for container ships, with 541 of the vessel class in its registry. Seven other Flag States had more than one hundred registered container ships. These are Liberia (415), Germany (248), Singapore (177), Cyprus (139), the Marshall Islands (118), and the United Kingdom (104). The Panamanian, Liberian, and Marshallese flags are open registries and considered by the International Transport Workers' Federation to be flags of convenience. By comparison, traditional maritime nations such as the United States and Japan only had 75 and 11 registered container ships, respectively.

Vessel purchases

In recent years, oversupply of container ship capacity has caused prices for new and used ships to fall. From 2008 to 2009, new container ship prices dropped by 19–33%, while prices for 10-year-old container ships dropped by 47–69%. In March 2010, the average price for a geared 500-TEU container ship was US$10 million, while gearless ships of 6,500 and 12,000 TEU averaged prices of US$74 million and US$105 million, respectively. At the same time, second-hand prices for 10-year-old geared container ships of 500, 2,500, and 3,500-TEU capacity averaged prices of US$4 million, US$15 million, and US$18 million, respectively. In 2009, 11,669,000 gross tonnes of newly built container ships were delivered. Over 85% of this new capacity was built in South Korea, China, and Japan, with South Korea accounting

Table 2.4 Worldwide capacity (1990–2017)

Year	*Capacity (TEUs, m)*
1990	1.5
2000	4.3
2008	10.6
2012	15.4
2017	20.3

for the lions share with over 57% of the world's total. New container ships accounted for 15% of total new tonnage, which is way behind bulk carriers at 28.9% and oil tankers at 22.6% (Table 2.4).

Scrapping

Most ships are removed from the fleet through a process known as scrapping. Scrapping is rare for ships under 18 years of age and common for those over 40 years of age. Shipowners and buyers negotiate scrap prices based on factors such as the ship's empty weight (called the light tonne displacement or LTD) and prices in the scrap metal market. Scrapping rates are volatile with the price per light tonne displacement swinging from a high of US$650 per LTD in mid-2008 to US$200 per LTD in early 2009, before increasing to US$400 per LTD in March 2010. As of 2021, over 96% of the world's container ship scrapping activity takes place in China, India, Bangladesh, and Pakistan. The global economic downturn of 2008–2009 resulted in more ships than usual being sold for scrap. In 2009, 364,300 TEU worth of container ship capacity was scrapped, up from 99,900 TEU in 2008. Container ships accounted for 22.6% of the total gross tonnage of ships scrapped for that year alone. Despite the surge, the capacity removed from the fleet only accounted for 3% of the world's container ship capacity. By 2021, the average age of container ships scrapped was 27.0 years (Figure 2.5).

Largest container ships

Economies of scale have dictated an upward trend in the size of container ships to reduce operating expenses. However, there are certain limitations to the size of container ships. Primarily, these are the availability of sufficiently large main engines and the availability of a sufficient number of ports and terminals prepared and equipped to manage ultra-large container ships. Furthermore, the permissible maximum ship dimensions in some of the world's main waterways present an upper limit in terms of vessel growth. This primarily concerns the Panama Canal, Suez Canal, English Channel and the Singapore Strait. That said, since even Very Large Container Ships are vessels with low draught compared with large tankers and bulk carriers,

Figure 2.5 Container ship *Unity* ready for scrapping, Bangladesh.

there is still considerable room for vessel growth. Compared with the current largest container ships, a 20,000–22,000 TEU container ship would only be moderately larger in terms of exterior dimensions than the first iteration of ultra large container ships, the *Emma Mærsk class* (15,200 TEU). According to a 2011 estimate, an ultra-large container ship of 20,250 TEU would measure 440 m × 59 m (1,444 ft × 194 ft), compared with 397.71 m × 56.40 m (1,304.8 ft × 185.0 ft) for the *Emma Mærsk class*, with an estimated dwt of circa 220,000 tonnes dwt. Although such a vessel might be near the upper limit for a Suez Canal passage, the so-called Malaccamax concept (for Straits of Malacca) does not apply for container ships, since the Malacca and Singapore Straits' draught limit of about 21 m (69 ft) is still above that of any conceivable container ship design. In the present market situation, main engines will not present as much of a limiting factor for vessel growth either. The steadily rising expense of fuel oil in the late 2010s prompted most container lines to adapt a slower, more economical voyage speed of about 21 knots, compared with earlier top speeds of 25 or more knots. Subsequently, newly built container ships can be fitted with a smaller main engine. The engine types fitted to ships of 14,000 TEU in the mid-2010s are sufficiently powerful enough to propel the ultra large container ships in operation today. Despite this, *Mærsk Line*, the world's largest container shipping company, opted for twin engines (two smaller engines working two separate propellers) when ordering a series of ten 18,000 TEU vessels from Daewoo Shipbuilding in February 2011. The ships were delivered between 2013 and 2014. In 2016, some experts believed that the current largest container ships are at the optimum size, and could not economically become any larger, as port facilities would be too expensive, port handling too time-consuming, the number of suitable ports too low, and insurance costs too high. In March 2017, the first ship with an official capacity over 20,000 TEUs, the *MOL Triumph*, was launched at the Samsung Heavy Industries. The *MOL Triumph* has a capacity of 20,150 TEUs. Between 2018 and 2021, a further 15 ultra large container vessels have been delivered with an official carrying capacity of between 20,119 and 23,992 TEUs (Tables 2.5, 2.6).

Freight market

The act of hiring a ship to carry cargo is called chartering. Outside special bulk cargo markets, ships are hired by three types of charter agreements: the voyage charter, the time charter, and the bareboat charter. In a voyage charter, the charterer rents the vessel from the loading port to the discharge port. With a time-charter, the vessel is hired for a set period, to perform voyages as the charterer directs. In a bareboat charter, the charterer acts as the ship's operator and manager, taking on responsibilities such as providing the crew and maintaining the vessel. The completed chartering contract is known as a charter party. UNCTAD tracks two aspects of container shipping prices in its annual Review of Maritime Trade. The first is a chartering price, specifically the Price to Time-charter one TEU slot for 14 tonnes of cargo on a container ship. The second is the freight rate, or comprehensive daily cost to deliver one-TEU worth of cargo on a given route. As a result of the late-2000s recession, both indicators showed sharp drops during 2008–2009, though the market has since shown signs of stabilisation between 2010 and 2020. UNCTAD uses the Hamburg Shipbrokers' Association (formally the Vereinigung Hamburger Schiffsmakler und Schiffsagenten e. V. or VHSS for short) as its main industry source for container ship freight prices. The VHSS maintains several indices of container ship charter prices. The oldest, which dates from 1998, is the Hamburg Index. This index considers time-charters on FC container ships controlled by Hamburg brokers. It is limited to charters of 3 months or more and is represented as the average daily cost in US$ for a one-TEU slot with a weight of fourteen tonnes. The Hamburg Index data is divided into 10 categories based primarily on the vessel carrying capacity. Two additional categories exist for small vessels of under 500 TEU that carry their own cargo cranes. In 2007, VHSS started another index, the New ConTex, which tracks similar data obtained from an international group of shipbrokers. The Hamburg Index shows some

Table 2.5 Largest container ships (2017–2021)

#	Built	Vessel name	*Length overall*		*Beam*		*Maximum*			
			(m)	*(ft)*	*(m)*	*(ft)*	*TEU*	*GT*	*Operator*	*Flag*
1	2021	*Ever Ace*	399.9	1312	61.5	202	23,992	235,579	Evergreen (Taiwan)	Panama
	2021	*Ever Act*	399.9	1312	61.5	202	23,992	235,579	Evergreen (Taiwan)	Panama
	2021	*Ever Aim*	399.9	1312	61.5	202	23,992	235,579	Evergreen (Taiwan)	Panama
	2021	*Ever Alp*	399.9	1312	61.5	202	23,992	235,579	Evergreen (Taiwan)	Panama
2	2020	*HMM Algeciras*	399.9	1312	61.0	200.1	23,964	228,283	HMM (South Korea)	Panama
	2020	*HMM Copenhagen*	399.9	1312	61.0	200.1	23,964	228,283	HMM (South Korea)	Panama
	2020	*HMM Dublin*	399.9	1312	61.0	200.1	23,964	228,283	HMM (South Korea)	Panama
	2020	*HMM Gdansk*	399.9	1312	61.0	200.1	23,964	228,283	HMM (South Korea)	Panama
	2020	*HMM Hamburg*	399.9	1312	61.0	200.1	23,964	228,283	HMM (South Korea)	Panama
	2020	*HMM Helsinki*	399.9	1312	61.0	200.1	23,964	228,283	HMM (South Korea)	Panama
	2020	*HMM Le Havre*	399.9	1312	61.0	200.1	23,964	228,283	HMM (South Korea)	Panama
3	2020	*HMM Oslo*	399.9	1312	61.5	202	23,820	232,311	HMM (South Korea)	Panama
	2020	*HMM Rotterdam*	399.9	1312	61.5	202	23,820	232,311	HMM (South Korea)	Panama
	2020	*HMM Southampton*	399.9	1312	61.5	202	23,820	232,311	HMM (South Korea)	Panama
	2020	*HMM Stockholm*	399.9	1312	61.5	202	23,820	232,311	HMM (South Korea)	Panama
	2020	*HMM St Petersburg*	399.9	1312	61.5	202	23,820	232,311	HMM (South Korea)	Panama
4	2019	*MSC Gülsün*	399.9	1312	61.5	202	23,756	232,618	MSC (Switzerland)	Panama
	2019	*MSC Samar*	399.9	1312	61.5	202	23,756	232,618	MSC (Switzerland)	Panama
	2019	*MSC Leni*	399.9	1312	61.5	202	23,756	232,618	MSC (Switzerland)	Panama
	2019	*MSC Mia*	399.9	1312	61.5	202	23,756	232,618	MSC (Switzerland)	Panama
	2019	*MSC Febe*	399.9	1312	61.5	202	23,756	232,618	MSC (Switzerland)	Panama
	2019	*MSC Ambra*	399.9	1312	61.5	202	23,756	232,618	MSC (Switzerland)	Panama

(Continued)

Table 2.5 (Continued) Largest container ships (2017–2021)

#	Built	Vessel name	*Length overall*		*Beam*		*Maximum*			
			(m)	*(ft)*	*(m)*	*(ft)*	*TEU*	*GT*	*Operator*	*Flag*
5	2019	*MSC Mina*	399.8	1312	61.0	200.1	23,656	228,741	MSC (Switzerland)	Panama
	2019	*MSC Isabella*	399.8	1312	61.0	200.1	23,656	228,741	MSC (Switzerland)	Panama
	2019	*MSC Arina*	399.8	1312	61.0	200.1	23,656	228,741	MSC (Switzerland)	Panama
	2019	*MSC Nela*	399.8	1312	61.0	200.1	23,656	228,741	MSC (Switzerland)	Panama
	2019	*MSC Sixin*	399.8	1312	61.0	200.1	23,656	228,741	MSC (Switzerland)	Panama
	2021	*MSC Apolline*	399.8	1312	61.0	200.1	23,656	228,741	MSC (Switzerland)	Panama
	2021	*MSC Amelia*	399.8	1312	61.0	200.1	23,656	228,741	MSC (Switzerland)	Panama
	2021	*MSC Diletta*	399.8	1312	61.0	200.1	23,656	228,741	MSC (Switzerland)	Panama
	2021	*MSC Michelle*	399.8	1312	61.0	200.1	23,656	228,741	MSC (Switzerland)	Panama
	2021	*MSC Allegra*	399.8	1312	61.0	200.1	23,656	228,741	MSC (Switzerland)	Panama
6	2020	*CMA CGM Jacques Saade*	399.9	1312	61.3	201	23,112	236,583	CMA CGM (France)	France
	2020	*CMA CGM Champs Elysées*	399.9	1312	61.3	201	23,112	236,583	CMA CGM (France)	France
	2020	*CMA CGM Palais Royal*	399.9	1312	61.3	201	23,112	236,583	CMA CGM (France)	France
	2020	*CMA CGM Louvre*	399.9	1312	61.3	201	23,112	236,583	CMA CGM (France)	France
	2021	*CMA CGM Rivoli*	399.9	1312	61.3	201	23,112	236,583	CMA CGM (France)	France
	2021	*CMA CGM Montmartre*	399.9	1312	61.3	201	23,112	236,583	CMA CGM (France)	France
	2021	*CMA CGM Concorde*	399.9	1312	61.3	201	23,112	236,583	CMA CGM (France)	France
	2021	*CMA CGM Trocadero*	399.9	1312	61.3	201	23,112	236,583	CMA CGM (France)	France
	2021	*CMA CGM Sorbonne*	399.9	1312	61.3	201	23,112	236,583	CMA CGM (France)	France

7	2017	*OOCL Hong Kong*	399.9	399.9	1312	58.8	193	21,413	210,890	OOCL (Hong Kong)
	2017	*OOCL Germany*	399.9	399.9	1312	58.8	193	21,413	210,890	OOCL (Hong Kong)
	2017	*OOCL Japan*	399.9	399.9	1312	58.8	193	21,413	210,890	OOCL (Hong Kong)
	2017	*OOCL United Kingdom*	399.9	399.9	1312	58.8	193	21,413	210,890	OOCL (Hong Kong)
	2017	*OOCL Scandinavia*	399.9	399.9	1312	58.8	193	21,413	210,890	OOCL (Hong Kong)
	2018	*OOCL Indonesia*	399.9	399.9	1312	58.8	193	21,413	210,890	OOCL (Hong Kong)
8	2018	*COSCO Shipping Universe*	400.0	1312.3	58.6	192	21,237	215,553	COSCO (China)	Hong Kong
	2018	*COSCO Shipping Nebula*	400.0	1312.3	58.6	192	21,237	215,553	COSCO (China)	Hong Kong
	2019	*COSCO Shipping Galaxy*	400.0	1312.3	58.6	192	21,237	215,553	COSCO (China)	Hong Kong
	2019	*COSCO Shipping Solar*	400.0	1312.3	58.6	192	21,237	215,553	COSCO (China)	Hong Kong
	2019	*COSCO Shipping Star*	400.0	1312.3	58.6	192	21,237	215,553	COSCO (China)	Hong Kong
	2019	*COSCO Shipping Planet*	400.0	1312.3	58.6	192	21,237	215,553	COSCO (China)	Hong Kong
9	2018	*CMA CGM Antoine de Saint Exupéry*	00.0	1312.3	59.0	193.6	20,954	219,277	CMA CGM (France)	France
	2018	*CMA CGM Jean Mermoz*	400.0	1312.3	59.0	193.6	20,954	219,277	CMA CGM (France)	Malta
	2018	*CMA CGM Louis Blériot*	400.0	1312.3	59.0	193.6	20,954	219,277	CMA CGM (France)	Malta

(*Continued*)

Table 2.5 (Continued) Largest container ships (2017–2021)

#	Built	Vessel name	Length overall		Beam		Maximum	GT	Operator	Flag
			(m)	(ft)	(m)	(ft)	TEU			
10	2017	*Madrid Mærsk*	399.0	1309.1	58.6	192	20,568	214,286	Maersk (Denmark)	Denmark
	2017	*Munich Mærsk*	399.0	1309.1	58.6	192	20,568	214,286	Maersk (Denmark)	Denmark
	2017	*Moscow Mærsk*	399.0	1309.1	58.6	192	20,568	214,286	Maersk (Denmark)	Denmark
	2017	*Milan Mærsk*	399.0	1309.1	58.6	192	20,568	214,286	Maersk (Denmark)	Denmark
	2017	*Monaco Mærsk*	399.0	1309.1	58.6	192	20,568	214,286	Maersk (Denmark)	Denmark
	2018	*Marseille Mærsk*	399.0	1309.1	58.6	192	20,568	214,286	Maersk (Denmark)	Denmark
	2018	*Manchester Mærsk*	399.0	1309.1	58.6	192	20,568	214,286	Maersk (Denmark)	Denmark
	2018	*Murcia Mærsk*	399.0	1309.1	58.6	192	20,568	214,286	Maersk (Denmark)	Denmark
	2018	*Manila Mærsk*	399.0	1309.1	58.6	192	20,568	214,286	Maersk (Denmark)	Denmark
	2018	*Mumbai Mærsk*	399.0	1309.1	58.6	192	20,568	214,286	Maersk (Denmark)	Denmark
	2019	*Maastricht Mærsk*	399.0	1309.1	58.6	192	20,568	214,286	Maersk (Denmark)	Denmark
11	2017	*MOL Truth*	400.0	1312.3	58.8	193	20,170	210,678	ONE (Japan)	Marshall Islands
	2018	*MOL Treasure*	399.0	1312.3	58.8	193	20,170	210,678	ONE (Japan)	Marshall Islands
12	2017	*MOL Triumph*	400.0	1312.3	58.8	193	20,170	210,678	ONE (Japan)	Marshall Islands
	2017	*MOL Trust*	400.0	1312.3	58.8	193	20,170	210,678	ONE (Japan)	Marshall Islands
	2017	*MOL Tribute*	400.0	1312.3	58.8	193	20,170	210,678	ONE (Japan)	Marshall Islands
	2017	*MOL Tradition*	400.0	1312.3	58.8	193	20,170	210,678	ONE (Japan)	Marshall Islands
13	2019	*Ever Glory*	400.0	1312.3	58.8	193	20,160	219,775	Evergreen (Taiwan)	Liberia
	2019	*Ever Govern*	400.0	1312.3	58.8	193	20,160	219,688	Evergreen (Taiwan)	Panama
	2019	*Ever Globe*	400.0	1312.3	58.8	193	20,160	219,688	Evergreen (Taiwan)	Panama
	2019	*Ever Greet*	400.0	1312.3	58.8	193	20,160	219,688	Evergreen (Taiwan)	Panama

14	2018	*Ever Golden*	400.0	1312.3	58.8	193	20,124	219,079	Evergreen (Taiwan)	Panama
	2018	*Ever Goods*	400.0	1312.3	58.8	193	20,124	219,079	Evergreen (Taiwan)	Panama
	2018	*Ever Genius*	400.0	1312.3	58.8	193	20,124	219,079	Evergreen (Taiwan)	Panama
	2018	*Ever Given*	400.0	1312.3	58.8	193	20,124	219,079	Evergreen (Taiwan)	Panama
	2018	*Ever Gifted*	400.0	1312.3	58.8	193	20,124	219,352	Evergreen (Taiwan)	Singapore
	2019	*Ever Grade*	400.0	1312.3	58.8	193	20,124	219,158	Evergreen (Taiwan)	Panama
	2019	*Ever Gentle*	400.0	1312.3	58.8	193	20,124	217,612	Evergreen (Taiwan)	Liberia
15	2018	*COSCO Shipping Taurus*	399.8	1312	58.7	193	20,119	194,864	COSCO (China)	Hong Kong
	2018	*COSCO Shipping Gemini*	399.9	1312	58.7	193	20,119	194,864	COSCO (China)	Hong Kong
	2018	*COSCO Shipping Virgo*	399.9	1312	58.7	193	20,119	194,864	COSCO (China)	Hong Kong
	2018	*COSCO Shipping Libra*	399.7	1311	58.7	193	20,119	194,864	COSCO (China)	Hong Kong
	2018	*COSCO Shipping Sagittarius*	399.7	1311	58.7	193	20,119	194,864	COSCO (China)	Hong Kong

Table 2.6 Ships on order (2022–2023)

Delivery	*Operator*	*Number*	*Maximum TEU*
2023	Mediterranean Shipping Company (MSC)	4 × Hudong-Zhonghua Shipbuilding 4 × Jiangnan Shipyard 2 × Yangzijiang Shipbuilding	24,232
2023	Ocean Network Express (ONE)	6 × Imabari Shipbuilding and Japan Marine United	24,000
2023	Seaspan ULC	2 × TBC	24,000
2022	Evergreen	6 × Samsung Heavy Industries	23,992
2022	Evergreen	4 × Jiangnan Shipyard 4 × Hudong-Zhonghua Shipbuilding	24,004
2023	Hapag Lloyd	12 × Daewoo Shipbuilding & Marine Engineering	23,500+
2023	OOCL	6 × Nantong COSCO KHI Ship Engineering 6 × Dalian COSCO KHI Ship Engineering	23,000

clear trends in recent chartering markets. First, rates were increasing from 2000 to 2005. From 2005 to 2008, rates slowly decreased, and in mid-2008 began a 'dramatic decline' of approximately 75%, which lasted until rates stabilised in April 2009. Rates have ranged from US$2.70 to US$35.40 in this period, with prices lower on larger ships. The most resilient sized vessel in this time were those from 200 to 300 TEU, a fact UNCTAD attributes to lack of competition in this sector. Overall, in 2010, these rates rebounded, but remained at half of their 2008 values. By 2011, the index showed signs of recovery for container shipping, and combined with increases in global capacity, indicating a positive outlook for the sector over the near future.

UNCTAD also tracks container freight rates. Freight rates are expressed as the total price in US$ for a shipper to transport one TEU worth of cargo along a given route. Data is given for the three main container liner routes: US – Asia, US – Europe, and Europe – Asia. Prices are typically different between the two legs of a voyage, for example the Asia – US rates have been significantly higher than the return US – Asia rates over recent years. Both the volume of container cargo and freight rates have dropped sharply. From 2009 to 2019, the freight rates on the US – Europe route were sturdiest, whereas the Asia to US route fell the most. Liner companies responded to their overcapacity in several ways. For example, in early 2009, some container lines dropped their freight rates to zero on the Asia to Europe route, charging shippers only a surcharge to cover operating costs. They decreased their overcapacity by lowering the ship's speed (a strategy called 'slow steaming') and by laying up ships. Slow steaming increased the length of the Europe to Asia routes to a record high of over 40 days. Another strategy used by some companies was to manipulate the market by publishing notices

Table 2.7 Container sector alliances

Alliance	*Partners*	*Ships*	*Weekly services*	*Ports*	*Port pairs*
Ocean Alliance	CMA CGM, COSCO Shipping, Evergreen	323	40	95	1,571
THE Alliance	Hapag-Lloyd, HMM, Ocean Network Express, Yang Ming	241	32	78	1,327
2M Alliance	Mærsk Line, Mediterranean Shipping Company	223	25	76	1,152

of rate increases in the press, and when a notice had been issued by one carrier, other carriers followed suit. Increasingly, the Trans-Siberian Railway (TSR) has become a more viable alternative to container ships on the Asia to Europe route. This railway can typically deliver containers in one-third to one half of the time of a typical sea voyage.

Container sector alliances

To control costs and maximise capacity utilisation on ever-larger ships, vessel sharing agreements, cooperative agreements, and slot-exchanges have become a growing feature of the maritime container shipping industry. Since 2015, 16 of the world's largest container shipping lines have consolidated their routes and services accounting for 95% of container cargo volumes moving in the dominant east-west trade routes. Within these alliances, carriers remain operationally independent, as they are forbidden by antitrust regulators in multiple jurisdictions from colluding on freight rates or capacity (Table 2.7).

CONTAINER PORTS

Container traffic through a port is often tracked in terms of twenty-foot equivalent units or TEU of throughput. As of 2019, the Port of Shanghai was the world's busiest container port, with 43,303,000 TEU handled. That year, 7 of the busiest 10 container ports were in the People's Republic of China, with Shanghai in first place, followed by Ningbo (third), Shenzhen (fourth), Guangzhou (fifth), Qingdao (seventh), Hong Kong (eight) and Tianjin (ninth). Rounding out the top 10 ports were Singapore (second), Busan in South Korea (sixth), and Rotterdam in the Netherlands (tenth). In total, the busiest 20 container ports handled 220,905,805 TEU in 2009, almost half of the global total estimated container traffic for that year (465,597,537 TEU) (Figure 2.6).

Figure 2.6 Tanger Med container terminal, Morocco.

LOSSES AND SAFETY ISSUES

It has been estimated that between 1990 and 2008, container ships lost 2,000 and 10,000 containers at sea, costing US$370 million. A survey for the 6 years from 2008 through 2013 estimated average losses of individual containers overboard were 546 per year, and average total losses including catastrophic events such as vessel sinkings or groundings at 1679 per year. Most go overboard on the open sea during storms, but there are some examples of whole ships being lost with their cargo. When containers are dropped, they immediately become an environmental threat as they become 'marine debris'. Once overboard, they fill with water and sink if the contents cannot hold air. If the container does not sink, it presents major problem for both ships and the marine environment. As container ships get larger and stacking becomes higher, the threat of containers toppling into the sea during heavy seas increases. This results from a phenomenon called 'parametric rolling', which uniquely affects only container ships and cause a ship to roll as much as 30–40 degrees during rough seas. This rolling motion creates a powerful torque on a 10-high stack of containers. This torque can easily snap the lashings and locks that keep the stack in place, resulting in containers tumbling overboard. The threat of piracy is a major concern to all shipping and can cost a container shipping company as much as US$100 million per year in longer routes and higher transit speeds. Piracy is a problem that is most prevalent off the coast of East Africa, and especially around the Horn, though international efforts to reduce maritime piracy

Figure 2.7 Container ship *Ever Given* stuck in the Suez Canal, Egypt – 24 March 2021.

has led to a steady decline in attacks, though the risk remains high. Other areas subject to maritime piracy include the Nigerian basin, Malacca Straits and the Persian Gulf. As container ships have continued to expand in length and width, the problems associated with vessels of this size have become increasingly apparent. On 23 March 2021, for instance, the *Ever Given* blocked the Suez Canal for several days after being blown off course by high winds. The *Ever Given* incident was estimated by Lloyds List to cause losses of US$400 million every hour based on westbound traffic revenues of US$5.1 billion per day, and eastbound traffic revenues of US$4.5 billion per day (Figure 2.7).

In this chapter, we have examined the main characteristics of container ships. In the next chapter, we will look at feeder ships.

Chapter 3

Feeder ships

Feeder vessels or feeder ships are medium-size freight ships. In general, a feeder designates a seagoing vessel with an average capacity of 300–1,000 TEU. Feeders collect shipping containers from different ports and transport them to central container terminals where they are loaded onto bigger vessels, or for further transport into the hub port's hinterland. In that way, the smaller vessels feed the larger liners, which carry thousands of containers. Over the years, feeder lines have been established by organisations transporting containers over a predefined route on a regular basis. Feeder ships are often run by companies that also specialise in short sea shipping. These companies not only ship freight to and from major ports like Rotterdam for further shipment, but also carry containers between smaller ports, for example, between terminals located on the north-west European seaboard and ports situated in the Baltic Sea. Coastal trading vessels, also known as coasters or skoots, are shallow-hulled ships used for trade between locations on the same island or continent. Their shallow hulls mean that they can get through reefs where deeper-hulled seagoing ships usually cannot. Coasters can load and unload cargo in shallow ports (Figure 3.1).

SHORT-SEA SHIPPING

The modern terms short-sea shipping, marine highway, and motorways of the sea, and the more historical terms coastal trade, coastal shipping, coasting trade, and coastwise trade, all encompass the movement of cargo and passengers by sea along a coast, without crossing an ocean. Throughout the European Union, short-sea shipping (or a translation thereof) is the term used by the European Commission. Alternatively, many English-speaking countries use the British terms coasting trade and coastwise trade. The United States maintained these terms from its colonial era, including for the domestic slave trade that shipped slaves by water from the Upper South to major markets around North America, especially New Orleans. The United States began regulating general coasting trade as early as 1793, with 'An act for enrolling and licensing ships and vessels to be employed in the

DOI: 10.1201/9781003342366-5

Figure 3.1 Typical coastal feeder ship *Wybelsum*.

coasting trade and fisheries, and for regulating the same', which passed Congress on 18 February that year. Over the years, it has been codified as Title 46 of the United States Code, chapter 551 (46 USC Ch. 551), 'Coastwise Trade'. Some short-sea ship vessels are small enough to travel inland on inland waterways. Short-sea shipping includes the movements of wet and dry bulk cargoes, containers, and passengers around the coast (e.g. from Lisbon to Rotterdam or from New Orleans to Philadelphia). Typical ship sizes range from 1000 dwt to 15,000 dwt with draughts ranging from around 3 to 6 m (10–20 ft). Typical (and mostly bulk) cargoes include grain, fertilisers, steel, coal, salt, stone, scrap, minerals, and oil products (such as diesel oil, kerosene, and aviation fuel), containers, and passengers.

Short-sea shipping around the world

Europe

In Europe, short-sea shipping is at the forefront of the European Union's transportation policy. It currently accounts for 40% of all freight moved throughout Europe. In the United States, short-sea shipping has yet to be used to the same extent as it is in Europe, although this is slowly changing. The main advantages promoted for this type of shipping are alleviation in road traffic congestion, decreased air pollution, and overall cost savings to the shipper. Shipping goods by ship (one 4,000 dwt vessel is equivalent to between 100–200 average articulated lorries) is more efficient and cost-effective than road transport (though the goods, if bound inland, have to be transferred and delivered by lorry or rail) and is much less prone to theft and damage. Forty per cent of all freight moved in Europe is classified as short-sea shipping, but the greater percentage of this cargo moves through Europe's heartland on rivers, and not oceans. Between 2010 and 2020, the term short-sea shipping has evolved in a broader sense to include point-to-point cargo movements on inland waterways as well as inland to ocean ports for shipment overseas. The contrasting terms deep-sea shipping, intercontinental shipping, and ocean shipping refer to maritime traffic that crosses oceans. Short-sea shipping is also distinct from inland navigation, for example, between two cities along a river. In Europe, the main hub of short-sea shipping is Rotterdam (the Netherlands), which is the largest European port, with Antwerp (Belgium) in second place and Hamburg (Germany) in third place. The Netherlands plays a significant role in this, having developed a hybrid vessel design able to navigate the sea as well as the Rhine into the Ruhrgebiet. The Dutch and Belgian main waterways (Maas, Waal, Amsterdam-Rhine Canal, and the Scheldt) locks and bridges have been adapted or built accordingly. Because of congestion in the larger ports, several smaller (container) ports have been developed. The same goes for the Rhine-ports such as Duisburg and Dortmund, both in Germany. The ports of Hamburg, Felixstowe (now the largest port in the UK), and Le Havre are also significant players in short-sea shipping. In the Netherlands, the sector has seen rapid growth, aided by a tax-enabled investment scheme. The traditional region for building 'coasters' is the province of Groningen, where most wharfs have side-laying ship slides. The current trend is to have bare hulls made with labour coming from Poland and or Romania, and then finishing the ship in the Netherlands.

North America

Cargo movements on the Great Lakes Waterway and St. Lawrence Seaway system are classified as short-sea shipping under the broader definition. St. Lawrence Seaway Management Corporation of Canada and its North American counterpart, the St. Lawrence Seaway Development Corporation, have for the past several years promoted this concept under its marketing umbrella 'Hwy H^2O'. The concept is intended to use existing capacity on the 2,300 mi (3,700 km) St. Lawrence–Great Lakes corridor in harmony with rail and truck modes to reduce overland congestion. Great Lakes Feeder Lines of Burlington, Ontario, Canada, was the first company to operate a 'fit for purpose', European-built short-sea shipping vessel, named *Dutch Runner*, which operated on the St. Lawrence Seaway under the Canadian Flag. During the winter of 2008–2009, she operated a weekly, fixed service between Halifax and St. Pierre et Miquelon, carrying roll-on/roll-off, breakbulk, containers, and refrigerated goods. This cargo was loaded and discharged using two 35 tonne cranes. Another Canadian firm, Hamilton-based

McKeil Marine, operates a fleet of tug-and-barge combinations, which ship commodities such as tar, fuels, aluminium ingots, and break-bulk cargoes. Along the St. Lawrence River, McKeil Marine transports aluminium ingots from a smelter in Quebec to destinations in Ohio, some 944 nm (1086 mi or 1519 km). One barge carries the equivalent of 22,040 tonne trucks. America's Marine Highway is a programme designed to promote inland and coastal shipping. In 2001, the Port of New York and New Jersey initiated its Port Inland Distribution Network (PIDN), a project designed to increase a network of inland points for shipping. The programme also sought to strengthen rail-port connections. In 2003, the first barge service to commence operation was from the Port of Salem (NJ) on the Delaware River to the Port of Albany-Rensselaer on the Hudson River (NY). This service remained in operation until 2006 when it ceased trading.

CABOTAGE

Cabotage is like short-sea shipping, except those goods or passengers are transported between two places in the same country by a transport operator from another country. It was originally applied to shipping along the coastal routes, port to port, but now it applies to aviation, railways, and road transport as well. Cabotage rights are the right of a company from one country to trade in another country. Cabotage laws apply to merchant ships in most countries that have a coastline to protect the domestic shipping industry from foreign competition, preserve domestically owned shipping infrastructure for national security purposes, and ensure safety in congested territorial waters. In the United Kingdom, cabotage is governed by the various Navigation Acts. Indonesia implemented a cabotage policy in 2005 after previously allowing foreign-owned vessels to operate freely within the country. In the Philippines, the Tariff and Customs Code of the Philippines (Republic Act No. 1937), which is also known as the Cabotage Law, restricts coastwise trade or the transport of passengers and goods within the country to vessels with Philippine registry and which must secure a coastwise license from the Philippines Maritime Industry Authority. After the passage of Foreign Ships Co-Loading Act or the Republic Act No. 10,668 in 2015, foreign vessels with cargo intended to be exported out of the country may dock in multiple ports in the country before transiting to a foreign port. China does not permit foreign flagged vessels to conduct domestic transport or domestic transshipments without the prior approval of the Ministry of Transport. While Hong Kong and Macau maintain distinct internal cabotage regimes from the mainland, maritime cabotage between territory and the mainland is considered domestic carriage and accordingly is off limits to foreign vessels. Similarly, maritime crossings across the Taiwan Strait require special permits from both the People's Republic of China and the Republic of China and are usually off-limits to foreign vessels.

In the European Union, rights to cabotage in newly admitted member States (in particular, Greece, Spain, and Portugal) were initially restricted though these provisions and were abandoned following criticisms after the Paros ferry disaster.[1] The Hague–Visby Rules, a convention which imposes duties on maritime carriers, apply only to 'carriage of goods by sea between ports in two different states', and thus do not apply to cabotage shipping. However, section 1(3) of the *UK Carriage of Goods by Sea Act (1971)* declares that the Rules 'shall have effect ... where the port of shipment is a port in the United Kingdom, whether or not the carriage is between ports in two different States ...'. In the United States, the *Merchant Marine Act of 1920* (the Jones Act) requires that all goods transported by water between United States ports be transported on ships that have been constructed in the United States and that fly the United States Flag, are owned by United States citizens, and are crewed by United States citizens and United States permanent residents. The *Passenger Vessel Services Act of 1886* states that no foreign vessels are permitted to transport passengers between ports or places in the United States, either directly or by way of a foreign port.

In this chapter, we have examined the role of feeder ships, and how they are an essential function for local and regional merchant shipping. In Chapter 4, we will look at general cargo ships.

NOTE

1 The *MS Express Samina* was a French-built ROPAX ferry that struck the charted Portes Islets rocks in the Bay of Parikia off the coast of Paros Island in the central Aegean Sea on 26 September 2000. The accident resulted in 82 deaths and the loss of the ship. The cause of the accident was found to be crew negligence, for which several members were held criminally liable. Witnesses allege that the crew was watching the match between Hamburg SV and Panathinaikos FC during the 2000–2001 UEFA Champions League instead of maintaining watch of the vessel.

Chapter 4

General cargo ships

In shipping, break-bulk cargo, also called general cargo, refers to goods that are stowed on board ship in individually counted units. Traditionally, the large numbers of items were recorded on distinct bills of lading that listed them by different commodities. This contrasts with cargo stowed in modern shipping containers as well as bulk cargo, which goes directly, unpackaged and in copious quantities, into a ship's hold(s), measured by volume or weight (for instance, oil or grain). The term break-bulk derives from the phrase breaking bulk – using 'to break-bulk' as a verb: to initiate the extraction of a portion of a ship's cargo, or to begin the unloading process from the ship's hold(s). Ships carrying break-bulk cargo are called general cargo ships. Break-bulk/general cargo consists of goods transported, stowed, and handled piecemeal, typically bundled somehow in unit loads for hoisting, either with cargo nets, slings, or crates, or stacked on trays, pallets, or skids. Furthermore, batches of break-bulk goods are frequently packaged in smaller containers, including bags, boxes, cartons, crates, drums, or barrels and vats. Traditionally, break-bulk cargo was lifted directly into and out of a vessel's holds, and this is still mostly the case today. Otherwise, it must be lifted onto and off its deck, by cranes or derricks present on the dock or on the ship itself. If hoisted on deck instead of straight into the hold, liftable or rolling goods must be man-handled and stowed competently by stevedores. Securing break-bulk and general freight inside a vessel includes the use of dunnage.[1] When no hoisting equipment is available, break-bulk has traditionally been manually carried on and off ship, over a plank, or it might be passed from man to man via a human chain. A break-in-bulk point is a place where goods are transferred from one mode of transport to another, for example the docks where goods transfer from ship to lorry or rail.

Break-bulk was the most usual form of cargo for most of the history of shipping. Since the late 1960s, the volume of break-bulk cargo has declined dramatically, relative to containerised cargo, while the latter has grown exponentially worldwide. Containerising makes cargo effectively more homogenous, like other bulk cargoes, and enables the improved economies of scale. Moving cargo on and off ship in containers is much more efficient, allowing ships to spend less time in port. Containerisation, once widely

DOI: 10.1201/9781003342366-6

Figure 4.1 Typical general cargo ship *Happy Dynamic*.

accepted, reduced shipping, and loading costs by as much as 80–90%. Break-bulk cargo also suffered significantly from theft and damage.

Although cargo of this sort can be delivered straight from a lorry or train onto the ship, the most common way is for the cargo to be delivered to the dock in advance of the arrival of the ship and for the cargo to be stored in warehouses. When the ship arrives, the cargo is taken from the warehouse to the quay and then lifted on board by either the ship's gear (derricks or cranes) or by quayside cranes. The discharge of the ship is the reverse of the loading operation. Loading and discharging by break-bulk is labour-intensive. The cargo is brought to the quay next to the ship and then each individual item must be lifted on board separately. Some items such as sacks or bags can be loaded in batches by using a sling or cargo net and others, such as cartons can be loaded onto trays before being lifted on board. Once on board, each item must be stowed separately. Before any loading takes place, any signs of the previous cargo are removed. The holds are swept, washed if necessary, and any damage to them repaired. Dunnage is laid ready for the cargo or is just put in bundles ready for the stevedores to lay out as the cargo is loaded.

There are a great many kinds of break-bulk cargo, including:

- Bagged cargo: Bagged cargo (e.g. coffee in sacks) is stowed on double dunnage and kept clear of the ship's sides and bulkheads. Bags are kept away from pillars and stanchions by covering it with matting or waterproof paper.
- Baled goods: Baled goods are stowed on a single dunnage at least 50 mm (2 in) thick. The bales must be clean with all the bands intact. Stained or oily bales are rejected. All fibres can absorb oil and are liable to spontaneous combustion. As a result, they are kept clear of

any new paintwork. Bales close to the deckhead are covered to prevent damage by dripping sweat.

- Barrels and casks: Wooden barrels are stowed on their sides on 'beds' of dunnage, which keeps the middle of the side (the bilge) off the deck. The barrel is stowed with the bung at the top. To prevent movement, wedges called quoins are put in on top of the 'beds'. Barrels should be stowed fore and aft and not athwart ships. Once the first tier has been loaded, the next tier of barrels fits into the hollows between the barrels; this is known as stowing 'bilge and cantline'. Barrels, which are also known as casks or tuns, are primarily used for transporting liquids such as wine, water, brandy, whisky, and even oil. They are usually built in a spherical shape to make them easier to roll and have less friction when changing direction.
- Corrugated boxes: Corrugated boxes are stowed on a thick layer of dunnage and kept clear of any moisture. Military and weather-resistant grades of corrugated fibreboard may be used. They are not over-stowed with anything other than similar boxes. They are frequently loaded on pallets to form a unit load; where so, the slings that are used to load the cargo are frequently left on to facilitate discharge.
- Wooden shipping containers: Wooden boxes or crates are stowed on double dunnage in the holds and single dunnage in the 'tween decks'. Heavy boxes are given bottom stowage. The loading slings are often left on to aid discharge.
- Drums: Metal drums are stowed on end with dunnage between tiers, in the longitudinal space of the ship.
- Paper reels: Reels or rolls are stowed on their sides and care is taken to make sure they are not crushed.
- Motor vehicles: Motor vehicles are lifted on board and then secured using lashings. Great care is taken to prevent damage. Vehicles are prepared by removing hazardous liquids such as petrol and diesel. This differs from the RORO vessels wherein vehicles are driven on and off the ship under their own power.
- Steel girders: Any long heavy items are stowed fore and aft. If they are stowed athwartship, they are liable to shift if the ship rolls heavily and could pierce through the side of the ship.

ADVANTAGES AND DISADVANTAGES

The biggest disadvantage with break-bulk is that it requires more port resources at both ends of a ship's journey, including longshoremen, loading cranes, warehouses, and transport vehicles. Moreover, they often assume more dock space due to multiple vessels carrying multiple loads of break-bulk cargo. Indeed, the decline of break-bulk did not start with containerisation; rather, the advent of tankers and bulk carriers reduced the need for transporting liquids in barrels and grains in sacks. Such tankers and carriers use specialised ships and shore facilities to deliver larger amounts of cargo to the dock and effect faster turnarounds with fewer personnel once the ship arrives; however, they do require large initial investments in ships, machinery, and training, slowing their spread to areas where investment to overhaul port operations and/or training for dock personnel in the handling of cargo on the newer vessels may not be available. As modernisation of ports and shipping fleets spreads across the world, the advantages of using containerisation and specialised ships over break-bulk have sped the overall decline of break-bulk operations. Alternatively, break-bulk continues to hold an advantage in areas where port development has not kept pace with shipping technology; break-bulk shipping requires minimal shore facilities – a quay for the ship to tie to, dock workers to assist in unloading, and warehouses to store materials for later reloading onto other forms of transport. As a result, there are still some areas of the world where break-bulk shipping continues to thrive. Goods shipped in break-bulk can also be offloaded onto smaller vessels and lighters for transport into even the most minimally developed ports in which the normally large container ships, tankers, and bulk carriers might not be able to access due to size and/or water depth. In addition, some ports capable of accepting larger container ships/tankers/bulk transporters still require goods to be offloaded in break-bulk fashion; for example, in the outlying islands of Tuvalu, fuel oil for the power stations is delivered in bulk but must be offloaded in barrels.

In this chapter, we have discussed some of the main characteristics of general cargo ships. Essentially, these types of vessels will carry any cargoes that can be loaded and unloaded by crane and are usually found in smaller ports where liners are less likely to be frequent. In the next chapter, we will be looking at reefer or refrigerated cargo ships.

NOTE

1 Dunnage for securing cargo in holds of ships has evolved from wooden boards forming 'cribs' to modern mechanical, spring-loaded post-and-socket systems, such as the 'pogo sticks' used on US Navy Combat Logistics Force (CLF) ships, which provide underway replenishment of stores, spares, repair parts, ammunition, ordnance, and liquids in cans and drums. Dunnage segregates cargo in the hold and prevents the cargo from shifting in response to the ship's motions.

Chapter 5

Reefer ships

Reefer ships are a type of refrigerated cargo ship typically used to transport perishable cargo, which require temperature-controlled handling, such as fruits, meat, vegetables, dairy products, and comparable items. Reefer ships are usually categorised into one of three types: side door vessels, conventional vessels, and refrigerated container ships. Side-door vessels have water-tight ports on the ship's hull, which open into a cargo hold. Elevators or ramps leading from the quay serve as loading and discharging access for forklifts or conveyors. Inside these access ports or side doors, pallet lifts or another series of conveyors bring the cargo to the respective decks. This distinctive design makes the vessels particularly well suited for inclement weather operations, as the tops of the cargo holds are always closed against rain and sun. Conventional vessels have a traditional cargo operation with top opening hatches and cranes/derricks. On such ships, when facing wet weather, the hatches need to be closed to prevent heavy rain from flooding the holds. Both aforementioned ship types are well suited for the handling of palletised and loose cargo. Refrigerated container ships are specifically designed to carry containerised unit loads where each container has its individual refrigerated unit. These containers are always 20-foot equivalent units that are the size of 'standard' cargo containers that are loaded and unloaded at container terminals and on-board container ships. These ships differ from conventional container ships in their design, power generation, and electrical distribution equipment. They need provisions made for powering each container's cooling system. Because of their ease of loading and unloading cargo, many container ships are now being built or redesigned to carry refrigerated containers (Figure 5.1).

Historically, a major use of refrigerated cargo hold type ships was for the transportation of bananas and frozen meat, but most of these ships have been partly replaced by refrigerated containers that have a refrigeration system attached to the rear end of the container. Whilst on the ship these containers are plugged into an electrical outlet (typically 440 V AC) that connects into the ship's power generation. Refrigerated container ships are not limited by the number of refrigeration containers they can carry, unlike other container ships, which may be limited in their number of refrigeration

DOI: 10.1201/9781003342366-7

Figure 5.1 Typical reefer ship *Baltic Melody*.

outlets or have insufficient generator capacity. Each reefer container unit is typically designed with a stand-alone electrical circuit and has its own breaker switch that allows it to be connected and disconnected as required. In principle, each individual unit could be repaired while the ship is still underway. Refrigerated cargo is a key revenue for some shipping companies. On multipurpose ships, refrigerated containers are mostly carried above the deck, as they must be checked for proper operation. Moreover, a major part of the refrigeration system (such as a compressor) may fail, which would have to be replaced or unplugged quickly in the event of a fire. Modern container vessels stow the reefer containers in cell guides with adjacent inspection walkways that enable reefer containers to be carried in the holds as well as on deck. Modern refrigerated container vessels are designed to incorporate a water-cooling system for containers stowed below deck. This does not replace the refrigeration system but facilitates cooling down of the external machinery. Containers stowed on the exposed upper deck are air-cooled, while those under deck are water-cooled. The water-cooling design allows capacity loads of refrigerated containers under deck, as it enables the dissipation of the high amount of heat they generate. This system draws fresh water from the ship's water supply, which in turn transfers the heat through heat exchangers and out into the open sea. There are also refrigeration systems that have two compressors for very precise and low-temperature operations, such as transporting a container of blood to war zones. Cargoes of shrimp, asparagus, caviar, and blood are considered among the most expensive refrigerated items. Bananas, fruit, and meat have historically been the main cargo of refrigerated ships.

DEVELOPMENT OF THE REEFER SHIP

In 1869, reefers were shipping beef carcasses frozen in a salt-ice mixture from Indianola, Texas, to New Orleans, Louisiana, to be served in hospitals, hotels, and restaurants. In 1874, shipping of frozen beef from America to London began, which developed into an annual trade of around 10,000 short tonnes (8,900 long tonnes; 9,100 tonnes[1]). The insulated cargo space was cooled by ice, which was loaded on departure. The success of this method was limited by insulation, loading techniques, ice block size, distance, and climate. The first attempt to ship refrigerated meat was made when the *Northam* sailed from Australia to the United Kingdom in 1876. The refrigeration machinery broke down on-route resulting in the loss of the cargo. In 1877, the steamers *Frigorifique* and *Paraguay* carried frozen mutton from Argentina to France, proving the concept of refrigerated ships, if not the economics. In 1879, *Strathleven*, equipped with compression refrigeration systems, successfully sailed from Sydney, Australia, to London, England, with 40 tonnes of frozen beef and mutton as a small part of her cargo. The clipper sailing ship *Dunedin*, owned by the New Zealand and Australian Land Company (NZALC), was refitted in 1881 with a Bell-Coleman compression refrigeration machine. This steam-powered freezer unit worked by compressing air, then releasing it into the hold of the ship. The expanding air absorbed heat as it expanded, cooling the cargo in the hold. Using three tonnes of coal a day, this steam powered machine could chill the hold to 22°C (40°F) below the surrounding air temperature, freezing the cargo in the temperate climate of southern New Zealand, and then maintaining it below freezing (0°C; 32°F) through the tropics. *Dunedin's* most visible sign of being an unusual ship was the funnel for the refrigeration plant placed between her fore and main masts (sometimes leading her to be mistaken for a steamship, which had been common since the 1840s). In February 1882, *Dunedin* sailed from Port Chalmers, New Zealand, with 4331 muttons, 598 lamb, and 22 pig carcasses, 246 kegs of butter, and hare, pheasant, turkey, chicken, and 2226 sheep tongues, arriving in London after sailing 98 days with its cargo still frozen. After meeting all costs, the NZALC made a £4,700 profit from the voyage. Soon after the *Dunedin's* successful voyage, an extensive frozen meat trade from New Zealand and Australia to the United Kingdom developed with over 16 different refrigerated and passenger refrigerated ships built or refitted by 1900 in Scotland and northern English shipyards. Within 5 years, 172 shipments of frozen meat were sent from New Zealand to the United

Kingdom. Refrigerated shipping also led to a broader meat and dairy boom in Australia, New Zealand, and Argentina, which remains the same today.

In 1880, the Nelson brothers of Ireland started shipping live beef from County Meath, Ireland to Liverpool, England. They successfully expanded their beef business until their imports from Ireland were insufficient to supply their rapidly growing business. Nelson decided to investigate the possibility of importing meat from Argentina. The first refrigerated ship they bought was the *Spindrift*, which they renamed in 1890 to the *SS Highland Scot*. A vessel of 3060 gross, the vessel was fitted with a primitive refrigerating plant, which operated on the cold-air system. Although not particularly technologically advanced, even for the time, the vessel became one of the pioneer vessels in the cross-Atlantic trade of refrigerated meat and other perishable commodities. Their regularly scheduled shipments and ships developed into the Nelson Line that was formed in 1880 for the meat trade from Argentina to the United Kingdom. Refrigeration made it possible to import meat from the United States, New Zealand, Argentina, and Australia. All their ships had a 'Highland' first name. Nelson Line began passenger services in 1910 between London, England, and Buenos Aires, Argentina, and in 1913 came under control of the Royal Mail Steam Packet Company. In 1932, Royal Mail Group collapsed, after which Royal Mail Lines, Ltd. was formed, and Nelson Lines merged into the new company.

The United Fruit Company has used some type of reefers, often combined with cruise ship passenger accommodations, since about 1889. Because their cargo mostly consisted of bananas, they were nicknamed the 'Banana Fleet' and their vessels, the eponymous 'banana boats'. As the ships were painted bright white, to reflect sunlight and reduce heat build-up in the holds, in 1910, the company's ships received the moniker the '"Great White Fleet'. Since bananas are light and the normal shipping route was to Central America and then back to various North American ports, these ships were often built as combination cargo ships and what are now called cruise ships to pay for more of their operating expenses. To avoid American shipping regulations and taxes, they were registered in six other countries, with very few now maintaining United States registry. In the 1980s, the United Fruit Company was taken over by Chiquita Brands International, which now owns the largest fleet of banana boats in the world, none of which sails under the United States Flag.

According to the UNCTAD, there were 53,973 registered merchant ships in the world in 2021, with a total gross tonnage of 2,116,401,000 dwt. Of this, 548 were designed as refrigerated cargo ships. Because of the proliferation of self-contained refrigerated container systems on container ships, there are many more ships than those designed for only refrigerated cargo that are also carrying some refrigerated cargo. As of 2021, the countries with the largest numbers of reefer ships in their registries are the world's two most prominent flags of convenience: Panama (212) and Liberia (109).

REFRIGERATED CARGO SYSTEMS

Reefer or refrigerated cargo can be shipped using several different systems, depending on the type of cargo, the class of vessel, the shipment time and distance, and the availability of different methods of containment.

Refrigerated containers

A refrigerated container or reefer is an intermodal container (shipping container) used in intermodal freight transport that is capable of refrigeration for the transportation of temperature-sensitive, perishable cargo such as fruits, vegetables, meat, and other comparable items. While a reefer will have an integral refrigeration unit, they rely on external power, from electrical power points ('reefer points') at a land-based site, a container ship, or on quay. When being transported by road on a trailer or on a railway wagon, they can be powered from diesel powered generators ('gen sets'), which attach to the container. Refrigerated containers are capable of controlling temperatures ranging from –65°C (–85°F) to 40°C (104°F). Some reefers are equipped with a water-cooling system, which can be used if the reefer is stored below deck on a vessel without adequate ventilation to remove the heat generated. Water cooling systems are more expensive than air current ventilation to remove heat from cargo holds, and the use of water-cooling systems is declining. Air cooling and water cooling are usually combined. The impact on society of reefer containers is vast, allowing consumers all over the world to enjoy fresh produce at any time of year, without any obvious sign of quality diminishment. Another refrigeration system sometimes used where the journey time is short is total loss refrigeration, in which frozen carbon dioxide ice (or sometimes liquid nitrogen) is used for cooling. The cryogenically frozen gas slowly evaporates, and thus cools the container. The container is cooled for as long as there is frozen gas available in the system. These have been used in railway cars for many years, providing up to 17 days of temperature regulation. Full-size intermodal containers equipped with these 'cryogenic' systems can maintain their temperature for the 30 days needed for sea transport. Since they do not require an external power supply, cryogenically refrigerated containers can be stored anywhere on any vessel that can accommodate 'dry' (i.e. un-refrigerated) ocean freight containers (Figure 5.2).

Valuable, temperature-sensitive, or hazardous cargo often require utmost system reliability. This type of reliability can only be achieved through the installation of a redundant refrigeration system. A redundant refrigeration system consists of integrated primary and back-up refrigeration units. If the primary unit malfunctions, the secondary unit automatically starts. To provide reliable power to the refrigeration units, these containers are often fitted with one or more diesel generator sets. Containers fitted with these systems may be required for transporting certain dangerous goods to comply with IMO regulations.

Figure 5.2 Reefer containers.

Insulated containers

Insulated shipping containers are a type of packaging used to ship temperature sensitive products such as foods, pharmaceuticals, organs, blood, biologic materials, vaccines, and chemicals. The term can also refer to insulated intermodal containers or insulated swap bodies. A variety of constructions have been developed. Typically, insulated shipping containers are constructed of a vacuum flask, similar to a 'thermos' bottle, fabricated thermal blankets or liners, moulded expanded polystyrene foam (EPS, Styrofoam), other moulded foams such as polyurethane, polyethylene, sheets of foamed plastics, vacuum insulated panels (VIPs), reflective materials such as metallised film, bubble wrap or other gas filled panels, and other packaging materials and structures. Some are designed for singular use, while others are returnable for reuse. Some insulated containers are decommissioned refrigeration units. Some empty containers are sent to the shipper disassembled or 'knocked down', assembled and used, then knocked down again for easier return shipment. Shipping containers are available for maintaining cryogenic temperatures, with the use of liquid nitrogen. Some carriers offer these as a specialised service. Insulated shipping containers are part of a comprehensive cold chain which controls and documents the temperature of a product through its entire distribution cycle. The containers may be used with a refrigerant or coolant such as block or cube ice, slurry ice, dry ice, gel or ice packs (often formulated for specific temperature ranges), and phase change materials (PCMs). Some products (such as frozen meat) have sufficient thermal mass to contribute to the temperature control and no excess coolant is required. A digital temperature data logger or time temperature indicator is often enclosed to monitor the temperature inside the container for its entire shipment. Labels and appropriate documentation (internal and external) are usually required.

Personnel throughout the cold chain need to be aware of the special handling and documentation required for some controlled shipments. With some regulated products, complete documentation is required. The use of 'off the shelf' insulated shipping containers does not necessarily guarantee proper performance; therefore, several factors need to be considered, including:

- the sensitivity of the product to temperatures (high and low) and to time at temperatures;
- the specific distribution system being used, that is the expected (and worst case) time and temperatures;
- regulatory requirements;
- the specific combination of packaging components and materials being used.

In specifying an insulated shipping container, the two primary characteristics of the material are its thermal conductivity or R-value, and its thickness. These two attributes help determine the resistance to heat transfer from the ambient environment into the payload space. The coolant material load temperature, quantity, latent heat, and sensible heat help determine the amount of heat the parcel can absorb while maintaining the desired control temperature. Combining the attributes from the insulator and coolant allows analysis of the expected duration of the insulated shipping container system. It is wise (and sometimes mandatory) to achieve formal verification of the performance of the insulated shipping container. Laboratory package testing might include ASTM D3103-07, Standard Test Method for Thermal Insulation Performance of Packages, ISTA Guide 5B: Focused Simulation Guide for Thermal Performance Testing of Temperature Controlled Transport Packaging, and others. In addition, validation of field performance (performance qualification) is extremely useful.

Environmental impact

Parcel to pallet-sized insulated shipping containers have historically been single-use products due to the low-cost material composition of EPS and water-based gel packs. The insulation material typically finds its way into landfill streams, as it is not readily recyclable in most American and European recycling processes. The development of reusable high-performance shipping containers has been shown to reduce packing waste by 95% while also contributing significant savings to other environmental pollutants.

In this chapter, we have looked at the basic features of reefer or refrigerated ships. These are ships which have been specially designed to carry chilled and or frozen cargoes. By keeping the cargo refrigerated, the quality and lifespan of the cargo is maintained. In the next chapter, we will turn our attention to roll on or roll off vessels.

NOTE

1 Long tonne, also known as the imperial tonne or displacement ton, is the name for the unit called the 'tonne' in the Imperial system of measurements (the avoirdupois system of weights). It was standardised in the 13th century. It is used in the United Kingdom and several other Commonwealth countries alongside the mass-based metric tonne defined in 1799, as well as in the United States for bulk commodities. It is not to be confused with the short tonne, a unit of weight equal to 2000 lbs (907.18474 kg) used in the United States, and in Canada before metrication.

Chapter 6

RORO vessels

RORO ships are cargo ships designed to carry wheeled cargo such as cars, trucks, semi-trailer trucks, buses, trailers, and railroad cars, which are driven on and off the ship on their own wheels or using a platform vehicle, such as a self-propelled modular transporter. This contrasts with LOLO vessels, which use a crane to load and unload cargo. RORO vessels have either built-in or shore-based ramps or ferry slips that allow the cargo to be efficiently rolled on and off the vessel when in port. While smaller ferries that operate across rivers and other short distances often have built-in ramps, the term RORO is reserved for large oceangoing vessels. The ramps and doors may be in the stern, bow, or sides, or a combination thereof. The RORO ship offers several advantages over traditional ship types. For the shipper, the number one benefit is speed. Since cars and lorries can drive straight into the ship at one port, and off again at the next port, usually within a few minutes of docking, this saves substantial time. It also integrates well with other cargo carrying methods such as containers. This has led to the development of hybrid vessels that carry both mobile cargoes and containers. Types of RORO vessels include ferries, cruise ferries, cargo ships, and barges. New automobiles that are transported by ship are often moved on a large type of RORO called a pure car carrier (PCC) or pure car/truck carrier (PCTC) (Figure 6.1).

Elsewhere in the shipping industry, cargo is normally measured by the tonne, but RORO cargo is typically measured by lanes in metres (LIMs). This is calculated by multiplying the cargo LIM by the number of decks and by its width in lanes. Lane widths differ from vessel to vessel, and there are several industry standards. The most used method for car ferries is a strip of deck 1 m (3.28 ft) long. A lane is conventionally 2 m (6.56 ft) wide, so that a lane metre is equivalent to 2 m^2 (21.528 ft^2). On ferries, the rule of thumb is that a car on a car ferry will need six lane metres, and a European semi-trailer 18 lane metres. On PCCs, cargo capacity is often measured in RT or RT43 units (based on a 1966 Toyota Corolla, the first mass-produced car to be shipped in specialised car-carriers and used as the basis for RORO vessel size. One RT is approximately 4 m (13.12 ft) of lane space required to store a 1.5 m (4.92 ft) wide Toyota Corolla) or in car-equivalent units (CEUs).

DOI: 10.1201/9781003342366-8

Figure 6.1 Typical roll on roll off ferry *P&O Pride of Hull*.

The largest RORO passenger ferry is the *MS Color Magic*, a 75,100 dwt cruise ferry that entered service in September 2007 for Color Line A/S of Norway. Built in Finland by Aker Finnyards, the vessel has a length overall of 223.70 m (733.11 ft) long and a beam of 35 m (114.10 ft), with a maximum capacity of 550 cars, or 1270 LIM of cargo. The RORO passenger ferry with the greatest car-carrying capacity is the *Ulysses* (named after the novel by James Joyce) and owned by Irish Ferries. *Ulysses* entered service on 25 March 2001 and operates between Dublin, Ireland, and Holyhead in Wales. The 50,938 metric-tonne ship has length overall of 209.02 m (685.9 ft), a beam of 31.84 m (104.6 ft), and a maximum carrying capacity of 1,342 cars over 4,101 lane metres of cargo (Table 6.1).

Table 6.1 Variations of different RORO vessels

Category	*Description*
ConRO	The ConRO (or RoCon) vessel is a hybrid of a RORO and a container ship. This type of vessel has a below-deck area used for vehicle storage while stacking containerised freight on the top decks. ConRo ships, such as the *G4 class* of Atlantic Container Line, can carry a combination of containers, heavy equipment, oversized cargo, cars, and commercial vehicles. A separate internal ramp system within the vessel segregates the cars and commercial vehicles from other vehicles, Mafi roll trailers, and break-bulk cargo.

(*Continued*)

Table 6.1 (*Continued*) Variations of different RORO vessels

Category	*Description*
LMSR	Large, Medium-Speed RORO (LMSR) refers to several classes of United States Military Sealift Command (MSC) RORO type cargo ships. Some are purpose-built to carry military cargo, while others are converted civilian vessels.
ROLO	A ROLO (roll-on/lift-off) vessel is another hybrid vessel type, with ramps serving vehicle decks but with other cargo decks only accessible when the tides change or by crane.
ROPAX	The acronym ROPAX (roll-on/roll-off passenger) describes a RORO vessel built for freight vehicle transport along with passenger accommodation. Technically, this encompasses all ferries with both a RORO car deck and passenger-carrying capacities, but in practice, ships with facilities for more than 500 passengers are often referred to as cruise ferries.

DEVELOPMENT OF THE RORO VESSEL

At first, wheeled vehicles carried as cargo on oceangoing ships were treated like any other cargo. Cars had their fuel tanks emptied and their batteries disconnected before being hoisted into the ship's hold, where they were chocked and secured. This process was tedious and difficult, and vehicles were subject to damage and could not be used for routine travel. The first known Roll On, Roll Off service was a train ferry, started in 1833 by the Monkland and Kirkintilloch Railway, which operated a wagon ferry on the Forth and Clyde Canal in Scotland. The first modern train ferry, the *Leviathan*, was built in 1849. The Edinburgh, Leith, and Newhaven Railway was formed in 1842 and the company wished to extend the East Coast Main Line further north to Dundee and Aberdeen. As bridge technology was not yet capable enough to provide adequate support for crossing over the Firth of Forth, which was five miles across, a different solution had to be found. The company hired the up-and-coming civil engineer Thomas Bouch who argued for a train ferry with a RORO mechanism to maximise the efficiency of the system. Ferries were to be custom-built, with railway lines and matching harbour facilities at both ends to allow the rolling stock to easily drive on and off. To compensate for the changing tides, adjustable ramps were positioned at the harbours and the gantry structure height was varied by moving it along the slipway. The wagons were loaded on and off with the use of stationary steam engines. Although other engineers had similar ideas, Bouch was the first to put them into effect, and did so with an attention to detail (such as design of the ferry slip), which led a subsequent President of the Institution of Civil Engineers to settle any dispute over priority of invention with the observation that 'there was little merit in a simple conception of this kind, compared with a work practically carried out in all its details, and brought to perfection'. The company was persuaded to install this train ferry service for the transportation of goods wagons across the

Firth of Forth from Burntisland in Fife to Granton. The ferry itself was built by Thomas Grainger, a partner of the firm Grainger and Miller. The service commenced on 3 February 1850 and was called *'the Floating Railway'*. The floating railway was initially intended as a temporary measure only until such time as the railway could build a bridge over the Firth of Forth. In fact, a bridge across the Forth was not completed until 1890, partly because of the catastrophic failure of Bouch's Tay Rail Bridge in 1879.

Train-ferry services were used extensively during World War I (1914–1918). From 10 February 1918, high volumes of railway rolling stock, artillery, and supplies for the Front were shipped to France from the 'secret port' of Richborough, near Sandwich on the south coast of England. This involved the construction of three train-ferries, each with four sets of railway lines on the main deck to allow for up to 54 railway wagons to be shunted directly on and off the ferry. These train-ferries could also be used to transport motor vehicles along with railway rolling stock. Later that month, a second train-ferry was established from the Port of Southampton on the south-east coast of England. In the first month of operations at Richborough, 5,000 long tonnes were transported across the English Channel. By the end of 1918, this had increased to 261,000 long tonnes. There were many advantages of the use of train-ferries over conventional shipping in World War I. It was much easier to move the large, heavy artillery and tanks that this kind of modern warfare required using train-ferries as opposed to repeated loading and unloading of cargo. With manufacturers loading tanks, guns, and other heavy munitions for shipping to the front directly on to wagons, which could be shunted on to a train-ferry in England and then shunted directly on

Figure 6.2 Typical roll on roll off container ship, *Rosa Delmas*, formerly *Rosa Tucano*, at La Rochelle, France.

to the French railway network, with direct connections to the front lines, many man hours of unnecessary labour were saved. An analysis conducted at the time found that to transport 1,000 long tonnes of war materiel from the point of manufacture to the front by conventional means involved the use of 1,500 labourers. When using train-ferries, that number decreased to around 100 labourers. This was of utmost importance, as by 1918, the British Railway companies were experiencing a severe shortage of labour with hundreds of thousands of skilled and unskilled labourers away fighting at the front. This meant economies and efficiency in transport had to be made wherever possible. After the signing of the Armistice on 11 November 1918, train ferries were used extensively for the return of materiel from the front. Indeed, according to British War Office statistics, a greater tonnage of materiel was transported by train ferry from Richborough in 1919 than in 1918. As the train ferries had space for motor transport as well as railway rolling stock, thousands of lorries, motor cars, and 'B Type' buses used these ferries to return from the continent.

During World War II (1939–1945), landing ships were the first purpose-built seagoing ships enabling road vehicles to roll directly on and off. The British evacuation at Dunkirk in 1940 demonstrated to the Admiralty that the Allies needed large, ocean-going ships capable of shore-to-shore delivery of tanks and other vehicles to facilitate amphibious assaults across the beaches of Europe. As an interim measure, three 4,000 and 4,800 tonne tankers, built to pass over the restrictive bars of Lake Maracaibo, Venezuela, were selected for conversion because of their shallow draught. Bow doors and ramps were added to these ships, which became the first tank landing ships. The first purpose-built LST design was *HMS Boxer*. It was a scaled down design from ideas penned by Winston Churchill. To carry 13 Churchill type infantry tanks, 27 vehicles and 200 men (in addition to the crew) at a speed of 18 knots, it could not have the shallow draught that would have made for easy unloading. As a result, each of the three vessels (*HMS Boxer, HMS Bruiser,* and *HMS Thruster*) ordered in March 1941 had a long ramp stowed behind the bow doors. In November 1941, a small delegation from the British Admiralty arrived in the United States to pool ideas with the United States Navy's Bureau of Ships regarding the development of ships and including the possibility of building further Boxer class vessels. During this meeting, it was decided that the Bureau of Ships would design these vessels. As with the standing agreement, these would be built by the Americans so British shipyards could concentrate on building vessels for the Royal Navy. The specification called for vessels capable of crossing the Atlantic and the original title given to them was 'Atlantic Tank Landing Craft' (Atlantic (TLC). Calling a vessel 91 m (300 ft) long a 'craft' was considered a misnomer and the type was re-christened 'Landing Ship, Tank (2)', or 'LST (2)'. The LST (2) design incorporated elements of the first British LCTs from their designer, Sir Rowland Baker, who was part of the British delegation. This included sufficient buoyancy in the ships' sidewalls that they would float even with the tank deck flooded. The LST (2) gave up the speed of *HMS Boxer* at only 10 knots (12 mph; 19 km/h) but had a similar load while drawing only 0.91 m (3 ft) forward when beaching. In three separate acts dated 6 February 1942, 26 May 1943, and 17 December 1943, the United States Congress provided the authority for the construction of LSTs along with a host of other auxiliaries, destroyer escorts, and assorted landing craft. The enormous building programme quickly gathered momentum. Such a high priority was assigned to the construction of LSTs that the previously laid keel of an aircraft carrier was hastily removed to make room for several LSTs to be built in her place. The keel of the first LST was laid down on 10 June 1942 at Newport News, Virginia, and the first standardised LSTs were floated out of their building dock in October. In total, 23 LST (2) class vessels were commissioned by the end of 1942.

At the end of World War I, vehicles were brought back from France to Richborough using the train ferry service. During the war, British servicemen recognised the enormous potential of landing ships and craft. The idea was simple: if you could drive tanks, munitions, and lorries directly onto a ship and then drive them off at the other end directly onto a beach, then theoretically you could use the same landing craft to conduct the same operation in the civilian commercial market, providing there were appropriate port facilities. From this idea grew the worldwide RORO industry of today. During the intervening period between the world wars, Lt. Colonel Frank Bustard formed the Atlantic Steam Navigation Company, with a view to commencing cheap transatlantic travel. Although this never materialised, during the war, he observed trials on Brighton Sands of an LST in 1943 when its peacetime capabilities were obvious. In spring 1946, the company approached the Admiralty with a request to purchase three of these vessels. The Admiralty were unwilling to sell, but after negotiations agreed to let the ASN have the use of three vessels on bareboat charter at a rate of £13 6s 8d per day. These vessels, *LSTs 3519*, *3534*, and *3512*, were renamed *Empire Baltic, Empire Cedric,* and *Empire Celtic*, respectively, perpetuating the name of the White Star Line ships in combination with the 'Empire' ship naming of vessels in government service during the war. On the morning of 11 September 1946, the first voyage of the Atlantic Steam Navigation Company took place when *Empire Baltic* sailed from Tilbury in England to Rotterdam, the Netherlands, with a full load of 64 vehicles for the Dutch Government. The original three LSTs were joined in 1948 by another vessel, LST 3041, renamed *Empire Doric*, after the ASN were able to convince commercial operators to support the new route between Preston, northern England, and the Port of Larne in Northern Ireland. The first sailing of this new route took place on 21 May 1948 by *Empire Cedric*. After the inaugural sailing, *Empire Cedric* continued the Northern Ireland service, offering initially a twice-weekly service. *Empire Cedric* was the first vessel of the ASN fleet to hold a passenger certificate and was allowed to carry 50 passengers. Thus, *Empire Cedric* became the first vessel in the world to operate as a commercial/passenger ROPAX ferry, and the ASN became the first commercial company to offer this type of service.

The first RORO service crossing the English Channel began from Dover in 1953. In 1954, the British Transport Commission (BTC) took over the ASN under the then Labour Government's nationalisation policy. In 1955, another two LSTs where chartered into the existing fleet, *Empire Cymric* and *Empire Nordic*, bringing the fleet strength to seven. The Hamburg service was terminated in 1955, and a new service was opened between Antwerp and Tilbury. The fleet of seven ships was to be split up with the usual three ships based at Tilbury and the others maintaining the Preston to Northern Ireland service. During late 1956, the entire fleet of ASN were taken over for use in the Mediterranean during the Suez Crisis, and the drive-on/drive-off services were not re-established until January 1957. At this point, ASN were made responsible for the management of 12 Admiralty LST (3)s brought out of reserve because of the Suez Crisis but were too late to see service. The first RORO vessel that was purpose-built to transport loaded semi-trailers was *Searoad of Hyannis*, which began operation in 1956. While modest in capacity, it could transport three semitrailers between Hyannis in Massachusetts and Nantucket Island. In 1957, the US military issued a contract to the Sun Shipbuilding and Dry Dock Company in Chester, Pennsylvania, for the construction of a new type of motorised vehicle carrier. The ship, *USNS Comet*, had a stern ramp as well as interior ramps, which allowed cars to drive directly from the dock, onto the ship, and into place. This method of loading and unloading was sped up dramatically. The *USNS Comet* also had an adjustable chocking system for locking cars onto the decks and a ventilation system to remove exhaust gases that accumulated during vehicle loading. During the 1982 Falklands War, the *SS Atlantic Conveyor* was requisitioned as an emergency aircraft and helicopter transport for British Hawker Siddeley Harrier STOVL fighter planes; one Harrier was kept fuelled, armed, and ready to VTOL launch for emergency air protection against long range Argentine aircraft. *SS Atlantic Conveyor* was sunk by Argentine Exocet missiles after offloading the Harriers to proper aircraft carriers, but the vehicles and helicopters still on board were lost.

The first cargo ships specially fitted for the transport of massive quantities of cars came into service in the early 1960s, when Volkswagen AG of Germany started exporting cars en masse to North America and Canada. During the 1970s, the market for exporting and importing cars had increased dramatically and the number and type of ROROs also increased. In 1970, Japan's K Line built the Toyota Maru No.10, Japan's first PCC, followed in 1973 by the European Highway, what was then the largest PCC at that time with a maximum capacity of 4200 vehicles. Today's PCCs and their sibling, the pure car truck carrier, are distinctive ships with a box-like superstructure running the entire length and breadth of the hull, fully enclosing the cargo.

They typically have a stern ramp and a side ramp for dual loading of thousands of vehicles (such as cars, trucks, heavy machineries, tracked units, Mafi roll trailers, and loose statics), and extensive automatic fire control systems. The PCTC has liftable decks to increase vertical clearance, as well as heavier decks for 'high-and-heavy' cargo. A 6,500 unit car ship, with 12 decks, can have three decks, which can take cargo up to 150 short tonnes (136 t; 134 long tonnes) with liftable panels to increase clearance from 1.7 to 6.7 m (from 5.7 to 22.0 ft) on some decks, albeit by lifting decks to accommodate higher cargo, reduces total carrying capacity. These vessels can achieve a cruising speed of 16 knots at eco-speed, while at full speed can achieve more than 19 knots. In 2007, the Swedish/Norwegian shipping company Wallenius Wilhelmsen launched the 8,000-car equivalent unit car carrier, *Faust*. This was the first of a new series of large car and truck carriers (LCTC). Currently, the largest car carrying vessels are the *Horizon class* owned by Norwegian company Höegh Autoliners, which have a capacity of 8500 CEU each. The car carrier *Auriga Leader*, belonging to Japan's Nippon Yusen Kaisha, built in 2008 with a capacity of 6,200 cars, was the world's first partially solar powered ship.

RORO STOWAGE AND SECURING OF CARGO

ROROs are often considered the workhorses of the sea, with their versatility to transport diverse cargoes and short port turnarounds. Given the fact that the cargo carrying capacity of any vessel determines the vessel's revenue potential, it is important to make optimal use of all available cargo space. With ROROs, this has become bit of a double-edged sword, with crews trying to match the unique characteristics of rolling cargo with static containerised cargo (e.g. awkward shapes, difficulties securing the cargo on deck, the lack of transverse bulkheads, loading conditions, and stability and rolling periods). To ensure cargoes are stowed correctly, and using the most optimum configuration possible, it is important to consider the shipper's special advice or guidelines regarding the handling and stowage of individual vehicles and cargo units. This typically means aligning vehicles in a fore to aft direction, and as closely together athwartships, so that in the event of a failure of the securing arrangements, the transverse movement of the cargo is restricted. This complex arrangement is further complicated by the need to maintain void spaces around the water spray fire curtains, operating controls for the bow and stern doors, entrances to accommodation spaces, ladders, stairways, companionways, access hatches, firefighting equipment, controls to deck scupper valves, and controls to the ventilation trunk fire dampers. Moreover, sufficient space must be kept between the front and rear of each vehicle to allow easy access for passengers and crew members. Parking brakes, where provided, must be applied fully. For semi-trailers, that is those trailers uncoupled from the prime mover, must not be supported on their landing legs during sea transport, unless the landing legs are specifically designed for use at sea and marked accordingly. The deck plates of the vessel must also be rated with sufficient strength to support the point loadings. Where the semi-trailer is not supported by their own landing legs, trestle supports, or similar devices should be placed in the immediate area of the drawplates so that the connection of the fifth wheel to the kingpin is not restricted. Depending on the area of operation, the predominant weather conditions, and the characteristics of the ship, freight vehicles should be stowed so that the chassis is kept as static as possible by not allowing free play in the suspension. This can be done by securing the vehicle to the deck as tightly as the lashing tensioning device permits or by jacking up the freight vehicle chassis prior to securing, or, in the case of compressed air suspension systems, by first releasing the air pressure where this facility is provided. Since compressed air suspension systems lose air, adequate arrangements should be made to prevent the slackening of lashings because of air leakage during the voyage. Such arrangements may include the jacking up of the vehicle or the release of air from the suspension system where this facility is provided. Proper securing of any cargo is of absolute importance to the safety of life at sea. The shape of RORO ships is such that any condition of instability can lead to disaster (Figure 6.3).

Figure 6.3 Car carrier *Atlas Leader*, arriving Casablanca, Morocco.

SAFETY ASPECTS

The IMO categorises ships according to their freeboard and internal subdivisions. Accordingly, there are two categories: Class A and Class B. Class A ships are those which have fewer sea openings and are thus better protected from the sea. They also have stringent subdivision restrictions. Class B ships, however, are those which have a higher freeboard and are governed by less stringent subdivision rules. Because of their unique design, RORO ships fall under the Class B categorisation. This is partly due to the presence of a large opening to the bow, stern or sides of the vessel, and the lack of integral transverse subdivision bulkheads from fore to aft. Whilst this might seem counter-intuitive from a design and safety perspective, it is necessary for the free flow of mobilised cargo onto and off the vessel. It is worth examining these design issues in a little more detail to understand the cause, justification, and consequences of the vessel design.

Lack of subdivisional bulkheads

As we have mentioned, RORO ships are designed and built without the standard transverse subdivisional bulkheads found in most other classifications and categories of ships. The problem with not having transverse subdivision bulkheads is an adverse one. Transverse bulkheads are incorporated into the ship's structure to maintain the damaged stability or watertight integrity of the ship in the event of any of the compartment's flooding. To a much lesser extent, bulkheads may also restrict the spread of fire. The bulkheads work by dividing the vessel into smaller compartments. Each compartment is sealed less for access gangways, companionways, and void spaces for pipes and cables. Should one or more compartments become flooded, the adjacent compartments should remain sealed and dry. Provided the ship remains stable, and no further compartments are affected, the ship will remain afloat. Without these transverse bulkheads, there is nothing to prevent the progressive flooding of the ship's hull. This not only causes the ship to lose inherent buoyancy, but also leads to the development of free surface effect. Free surface effect is not limited to RORO vessels, as we noted in the chapter on bulk carriers, but it is fair to suggest that RORO vessels are more prone, and therefore more likely to be affected by, free surface effect caused by progressive flooding (Figure 6.4).

Maintaining stability

Every RORO ship, being a Class B type of vessel, has considerable freeboard, which means it operates at a low draught. RORO ships have multiple tiers of decks for the stowage of different cargoes. Cars and commercial vehicles tend to be layered on hoistable or adjustable decks, with trailers and other large or out of gauge cargoes stowed under and around. This requires higher

Figure 6.4 Inside the hull of a typical RORO vessel.

overhead clearances. Because of this, the depth of these ships tends to be extremely high, owing to a high depth to draught ratio. Cargo is stowed up to the top-most deck, resulting in the rise in the accommodation deck. As a result of the increased depth to draught ratio, these ships are extremely sensitive to heeling moments. A heeling moment might not only be created by wind gusts or waves, but also by the shifting of internal cargo. Thus, cargo latching and locking systems must be regularly checked and ensured to prevent cargo shifts during voyages. Furthermore, heeling moments in lightship conditions can be prevented by incorporating heeling tanks at the port and starboard sides. Sadly, there have been many cases of RORO ships capsizing due to rapid heeling moments, giving the crew extraordinarily little time for proper evacuation. The sinking of the *MV Sewol* in Korea, 2014, and the *MS Express Samina* (2000), are examples of just how much RORO ships are at the mercy of the sea.

Vessel stiffness

Though this issue is related to stability, and is quite a complex and technical subject, it is interesting to discuss, nonetheless. The steel structure of RORO ships is designed to have an exceptionally low centre of gravity, as cargo is loaded up to the top-most deck. This offsets the rise in the ship's centre of gravity. But due to the risk of rapid heeling, the overall centre of gravity of

RORO ships must be kept low. Though doing so improves the ship's stability, it creates a problem itself. A reduced centre of gravity will always tend to decrease the rolling period of the ship. This means it will feel as if the ship is rolling too fast. Ships are designed, by the nature of their bottom hulls and ballast conditions, to return to their upright position. Ideally, this should happen rapidly but not so rapidly as to roll the vessel to the opposite side. Incidentally, it is this rolling motion that causes seasickness. When a ship rolls rapidly from one side to its upright centre, it is said to be stiff. Stiffness places high stresses on cargo lashing systems, which can cause them fail. When the lashing snaps or splits, rolling cargo can begin to shift leading the vessel to lose stability.

The cargo doors

Other than doors on the port and starboard sides, RORO ships also have aft or bow doors fitted with ramps. This facilitates the loading and unloading of heavier cargoes and rolling stock. Both these door types highlight the current problems in terms of vessel safety.

The stern door

By necessity of design, stern cargo doors are positioned close to the waterline of the ship. Where the stern door has not been properly secured, which is tantamount to gross human error as improper locking as the bridge and crew are notified by the ship's systems, as the ship transits out to sea, the improperly locked stern door may serve as the source of water ingress. Though this is a consequence of human error, several efforts to alter the design of these doors have been made, though it is impossible to position the stern door any higher above the waterline than is currently the case, as this would inhibit the operation of the stern ramp and the easy free flow of incoming and outgoing cargo (Figure 6.5).

The bow door

Many types of RORO ships are fitted with bow doors wherein the bow of the ship is itself a hydraulically hinged structure, which acts as a door, and out from which a ramp extends for cargo flow in and out of the ship. The bow of the ship is especially vulnerable to the impact of waves as the ship surges forward. This impact carries on continuously. Within the required time, the bow can begin to exhibit extreme stress, and without attention, will begin warp and fail. When this happens, it is a matter of time before the bow door is disgorged. One of the worst maritime accidents to occur in the history of merchant shipping involved this design of ship. The *MV Estonia* sank in the Baltic Sea on 28 September 1994 with the loss of 852 passengers and crew. Only 137 survivors made it ashore out of a full passenger and

Figure 6.5 Stern ramp of the *Mærsk Willow*.

crew manifest of 989 people. In the incident involving the *MV Estonia*, the bow door securing mechanism failed through a combination of poor maintenance and metal fatigue. Between 00:55 and 01:50, a series of particularly large frontal waves ripped the bow door off the vessel leaving the car decks hopelessly exposed. As the ship was fully laden, departing the Port of Tallin with a slight list towards starboard, caused by poor cargo distribution, there was truly little the crew could have done to save the stricken ship. Although the main cause of the *MV Estonia's* sinking was progressive down flooding caused by the loss of the bow door, the effect of 'rapid heeling' prevented the crew from evacuating the passengers quickly enough (Figure 6.6, Table 6.2).

Location of lifeboats

This is a matter of concern for all vessels, but especially so for RORO passenger ships. As we know that RORO vessels have a high freeboard, it is important to note the risk attached to such a design. In the event of rapid sinking, there have been cases where the lifeboats could not be successfully deployed from the embarkation deck due to its height above the waterline. It is due to this risk that most modern ROPAX ships are now equipped with inflatable chutes, which help passengers slide down from the embarkation deck, whenever the deployment of lifeboats is impeded (Figure 6.7).

Commercially, RORO ships have been one of the most successful due to their flexibility, integration into the global logistic chain, and operational speed. Yet, despite their commercial success, ROROs have always received

Figure 6.6 Bow door on the *GNV Cristal.*

Table 6.2 Passengers and crew of the *MV Estonia*

Nationalities	*Deaths*	*Survivors*	*Total*
Sweden	501	51	5,525
Estonia	285	62	347
Latvia	23	6	29
Russia	11	4	15
Finland	10	3	13
Norway	6	3	9
Germany	5	3	8
Denmark	5	1	6
Lithuania	3	1	4
Morocco	2	0	2
Netherlands	1	1	2
Ukraine	1	1	2
United Kingdom	1	1	2
Belarus	1	0	1
Canada	1	0	1
France	1	0	1
Nigeria	1	0	1
Total	852	137	989

Most passengers on board the *MV Estonia* were Swedish, although some were of Estonian origin. Most of the crew, including the master, Captain Arvo Andresson, were Estonian. Captain Andresson did not survive.

Figure 6.7 Typical placement of lifeboats on the ROPAX vessel *MS Lubeck Link*, Travemünde, Germany.

criticism for their design. In fact, it is those design factors which give ROROs their success is also in many respects the reason they can so spectacularly fail. The lack of internal bulkheads, a large cargo access door, poor stability, low freeboards, poor cargo stowage and securing practices, poor crew training, and poor provision and maintenance of lifesaving equipment are some of the major contributors to RORO-related incidents.

In the next chapter, we will look at fishing vessels.

Chapter 7

Fishing vessels

A fishing vessel is a boat or ship used to catch fish in the sea, or on a lake or river. Many kinds of vessels are used in commercial, artisanal, and recreational fishing. The total number of fishing vessels in the world in 2016 was estimated to be about 4.6 million, unchanged from 2006. The fishing fleet in Asia is by far the largest, consisting of 3.5 million vessels, accounting for 75% of the global fleet. In Africa and North America, the estimated number of vessels declined from 2014 by just over 30,000 and by 5,000, respectively. For Asia, Latin America, and the Caribbean and Oceania, the numbers have all increased, largely because of improvements in estimation procedures. It is difficult to estimate the number of recreational fishing boats, as they range in size from small dinghies to large charter cruisers, and unlike commercial fishing vessels, are often not dedicated just to fishing. Prior to the 1950s, there was little standardisation of fishing boats. Designs varied between ports and boatyards. Traditionally, the boats were built of wood, but wood is not often used now because of higher maintenance costs and lower durability. Fibreglass is used increasingly in smaller fishing vessels up to 25 m (82 ft), whilst steel is usually used on vessels above 25 m (82 ft) (Figure 7.1).

DEVELOPMENT OF FISHING BOATS

Early fishing vessels included rafts, dugout canoes, and boats constructed from a frame covered with animal hide or tree bark, like a coracle. The oldest boats found by archaeological excavation are dugout canoes dating back to the Neolithic Period around 7,000–9,000 years ago. These canoes were often cut from coniferous tree logs, using simple stone tools. A 7,000-year-old seagoing boat made from reeds and tar was found in Kuwait. These early vessels had limited capability; they could float and move on water but were not suitable for use in any great distance from the shoreline. The development of fishing boats took place in parallel with the development of boats for trade and war. Early navigators began to use animal skins or woven fabrics for sails. Affixed to a pole set upright in the boat, these sails gave

DOI: 10.1201/9781003342366-9

Figure 7.1 Traditional river boat used for fishing, Bon River, Vietnam.

early boats more range, allowing voyages of exploration. Around 4,000 BC, the Egyptians were building long narrow boats powered by many oarsmen. Over the next 1,000 years, they made a series of remarkable advances in boat design. They developed cotton-made sails to help their boats go faster with less work. Then, they built boats large enough to cross the oceans. These boats had sails and oarsmen and were used for travel and trade. By 3,000 BC, the Egyptians knew how to assemble planks of wood into a ship hull. They used woven straps to lash planks together, and reeds or grass stuffed between the planks to seal the seams. An example of their skill is the Khufu ship, a vessel 44 m (143 ft) in length entombed at the foot of the Great Pyramid of Giza around 2,500 BC and found intact in 1954.

At about the same time, the Scandinavians were also building innovative boats. People living near Kongens Lyngby in Denmark devised the idea of segregated hull compartments, which allowed the size of boats to gradually increase. A crew of some two dozen paddled the wooden *Hjortspring* boat across the Baltic Sea long before the rise of the Roman Empire. Scandinavians continued to develop better ships, incorporating iron and other metal into the design, and developing oars for propulsion. By 1000 AD, the Norsemen were pre-eminent on the oceans. They were skilled seamen and boat builders, with clinker-built boat designs that varied according to the type of boat. Trading boats, such as the knarrs, were wide to allow large cargo storage. Raiding boats, such as the longship, were long and narrow and extremely fast. The vessels they used for fishing were scaled-down versions of their cargo boats. The Scandinavian innovations influenced fishing boat design long after the Viking period ended. For example, yoles from the Orkney Island of Stroma were built in the same way as the Norse boats.

In the 15th century, the Dutch developed a type of seagoing herring drifter that became a blueprint for European fishing boats. This was the Herring Buss, used by Dutch herring fishermen until the early 19th centuries. The ship-type buss has a long history. It was known around 1,000 AD in Scandinavia as a bǘza, a robust variant of the Viking longship. The first herring buss was built in Hoorn around 1415. The ship was about 20 m (65 ft) long and displaced between 60 and 100 tonnes. It was a massive round-bilged keel ship with a bluff bow and stern, the latter high, and with a gallery. The busses used long drifting gill nets to catch the herring. The nets would be retrieved at night and the crews of 18–30 men would set to gibbing, salting, and barrelling the catch on the broad deck. During the 17th century, the British developed the dogger, an early type of sailing trawler or long liner, which commonly operated in the North Sea. Doggers were slow but sturdy, capable of fishing in the rough conditions of the North Sea. Like the herring buss, they were wide-beamed and bluff-bowed, but smaller, about 15 m (49.2 ft) long, had a maximum beam of 4.5 m (14.7 ft), a draught of 1.5 m (5 ft), and displacing about 13 tonnes. They could carry a tonne of bait, three tonnes of salt, half a tonne each of food and firewood for the crew and return with six tonnes of fish. Decked areas forward and aft provided accommodation, storage, and a cooking area. An anchor would have allowed extended periods of fishing in the same spot, in waters up to 18 m (59 ft) deep. The dogger would also have carried a small open boat for maintaining lines and rowing ashore.

A precursor to the dory type was the early French bateau, a flat bottom boat with straight sides used as early as 1671 on the St. Lawrence River. The common coastal boat of the time was the wherry and the merging of the wherry design with the simplified flat bottom of the bateau resulted in the birth of the dory. England, France, Italy, and Belgium have small boats from medieval periods that could be construed as predecessors of the Dory. Dories appeared in New England fishing towns sometime after the early 18th century. They were small, shallow-draft boats, usually about 5–7 m (15–22 ft) long. Lightweight and versatile, with high sides, a flat bottom and sharp bows, they were easy and cheap to build. The Banks dories appeared in the 1830s. They were designed to be transported on mother ships and used for fishing cod at the Grand Banks. Adapted directly from the low freeboard, French river bateaus, with their straight sides and removable thwarts, bank dories could be nested inside each other and stored on the decks of fishing schooners, such as the *Gazela Primeiro*, for their trip to the Grand Banks fishing grounds.

By the early 19th century, fishermen at Brixham in England needed to expand their fishing area further than ever before due to the ongoing depletion of stocks that was occurring in the overfished waters of South Devon. The Brixham trawler that evolved there was of a sleek build and had a tall gaff rig, which gave the vessel sufficient speed to make long-distance trips out to the fishing grounds in the ocean. They were also sufficiently robust to be able to tow large trawls in deep water. The great trawling fleet that built up at Brixham, earned the village the title of 'Mother of Deep-Sea Fisheries'. This revolutionary design made large-scale trawling in the ocean possible for the first time, resulting in a massive migration of fishermen from the ports in the South of England, to villages further north, such as Scarborough, Hull, Grimsby, Harwich, and Yarmouth, which were points of access to the large fishing grounds in the Atlantic Ocean. The small village of Grimsby grew to become the largest fishing port in the world by the mid-19th century. With the tremendous expansion in the fishing industry, the Grimsby Dock Company was formed in 1846. The dock covered 25 acres (10 ha) and was formally opened by Queen Victoria in 1854 as the first modern fishing port. The facilities incorporated many innovations of the time; the dock gates and cranes were operated by hydraulic power, and the 91 m (300 ft) Grimsby Dock Tower was built to provide a head of water with sufficient pressure by William Armstrong.

The elegant Brixham trawler spread across the world, influencing fishing fleets everywhere. Their distinctive sails inspired the song *Red Sails in the Sunset*, written aboard a Brixham sailing trawler called the *Torbay Lass*. By the end of the 19th century, there were over 3,000 fishing trawlers in commission in Britain, with almost 1,000 at Grimsby. These trawlers were sold to fishermen around Europe, including from the Netherlands and Scandinavia. Twelve trawlers went on to form the nucleus of the German fishing fleet. Although fishing vessel design increasingly began to converge around the world, local conditions still often led the development of diverse types of fishing boats. The Lancashire nobby was used down the north-west coast of England as a shrimp trawler from 1840 until World War II. The Manx nobby was used around the Isle of Man as a herring drifter. The fifie was also used as a herring drifter along the east coast of Scotland from the 1850s until well into the 20th century.

The earliest steam powered fishing boats first appeared in the 1870s and used the trawl system of fishing as well as lines and drift nets. These were large boats, usually 24–27 m (80–90 ft) in length with a beam of around 6.1 m (20 ft). They weighed 40–50 tonnes and travelled at 9–11 knots (10–13 mph; 17–20 km/h). The earliest purpose-built fishing vessels were designed and made by David Allan in Leith, Scotland, in March 1875, when he converted a drifter to steam power. In 1877, he built the first screw propelled steam trawler in the world. This vessel was named *Pioneer*. She was constructed of wood with two masts and carried a gaff rigged main and mizen using booms, and a single foresail. In 1878, he completed *Forward* and *Onward*, two steam-powered trawlers. Allan built a total of 10 boats at Leith between 1877 and 1881. Twenty-one boats were completed at Granton, his last vessel being *Degrave* in 1886. Most of these were sold to foreign owners in France, Belgium, Spain, and the West Indies. The first steamboats were made of wood, but steel hulls were soon introduced and were divided into watertight compartments. Steam trawlers were first introduced at Grimsby and Hull in the 1880s, although the steam drifter was not used in the herring fishery until 1897. By 1890, it was estimated that there were 20,000 men on the North Sea. The last sailing fishing trawler was built in 1925 in Grimsby. They were well designed for the crew with a large building that contained the wheelhouse and the deckhouse. The boats built in the 20th century only had a mizzen sail, which was used to help steady the boat when its nets were out. The main function of the mast was now as a crane for lifting the catch ashore. It also had a steam capstan on the foredeck near the mast for hauling nets. The boats had narrow, high funnels so that the steam and thick coal smoke was released high above the deck and away from the fishermen. These funnels were nicknamed woodbines because they looked like the popular brand of cigarette. These boats had a crew of 12 made up of a skipper, driver, fireman (to look after the boiler), and nine deck hands. Steam fishing boats had many advantages. They were usually about 6.1 m (20 ft) than the sailing vessels so that they could carry more nets and catch more fish. This was important, as the market was growing quickly at the beginning of the 20th century. They could travel faster and further and with greater freedom from weather, wind, and tide. Because less time was spent travelling to and from the fishing grounds, more time could be spent fishing. The steamboats also gained the highest prices for their fish, as they could return quickly to harbour with their fresh catch. The main disadvantage of the steamboats, though, was their high operating costs. Their engines were mechanically inefficient, while fuel and fitting out costs were extremely high. Before the First World War, building costs ranged anywhere between £3,000 and £4,000, at least three times the cost of the sail boats. To cover these prohibitive costs, they needed to fish for longer seasons. The higher expenses meant that more steam drifters were company-owned or jointly owned. As the herring fishing industry declined, steamboats became too expensive.

Trawler designs adapted as the way they were powered changed from sail to coal-fired steam by World War I to diesel and turbines by the end of World War II. The first trawlers fished over the side, rather than over the stern. In 1947, the company Christian Salvesen, based in Leith, Scotland, refitted a surplus *Algerine-class* minesweeper (*HMS Felicity*) with refrigeration equipment and a factory ship stern ramp, to produce the first combined freezer/stern trawler in 1947. The first purpose-built stern trawler was *Fairtry* built in 1953 at Aberdeen. The ship was much larger than any other trawlers then in operation and inaugurated the era of the 'super trawler'. As the ship pulled its nets over the stern, it could lift out a much greater haul of up to 60 tonnes. The *Lord Nelson* followed in 1961, installed with vertical plate freezers that had been researched and built at the Torry Research

Station in Aberdeen. These ships served as a basis for the expansion of 'super trawlers' around the world in the following decades. In recent years, commercial fishing vessels have been increasingly equipped with electronic aids, such as radio navigation aids and fish finders. During the Cold War, some countries fitted fishing trawlers with additional electronic gear so that they could be used as spy ships to monitor the activities of other countries.

Globally, the number of engine-powered vessels has been estimated to be 2.8 million (2016), which represents 61% of all fishing vessels, down from 64% in 2014, as the number of non-motorised vessels has increased, probably because of improved estimations. In 2016, about 86% of the motorised fishing vessels in the world were in the length overall class of less than 12 m (39 ft), the vast majority of which were undecked, and those small vessels dominated in all regions. Conversely, the largest vessels, classified as those with a length overall greater than 24 m (78 ft), made up about 2% of the total fleet. About 1.3 million of these are decked vessels with enclosed areas. All these decked vessels are mechanised, and 40,000 of them are over 100 tonnes. At the other extreme, two-thirds (approximately 1.8 million) of the undecked boats are traditional craft of several types, powered only by sail and oars. Artisan fishers use these boats (Figure 7.2).

Figure 7.2 Inshore fisheries trawler *William of Ladram*.

COMMERCIAL AND INDUSTRIAL FISHING VESSELS

As there is no internationally recognised fishing limit, coastal States may claim an Exclusive Economic Zone (EEZ), which may exceed up to 200 nautical miles and in which fishing may be carried out, if a country has claim to an EEZ, since this is not an automatic right. Sometimes, when States are semi-adjacent, they may need to come to an agreement as to the extent of their EEZs. The EEZ is referred to under articles 55 to 75 of the 1982 *UN Convention on the Law of the Sea* (UNCLOS). This specification of fishing rights has changed fishing patterns, and, in recent times, fishing boats have become more specialised and standardised. In the United States and Canada, large factory trawlers are extensively used, while the huge blue water fleets operated by Japan and the former Soviet-bloc countries have contracted. In Western Europe, fishing vessel design is focused on compact boats with high catching power. Commercial fishing is a high-risk industry, and countries are introducing regulations governing the construction and operation of fishing vessels. The IMO[1,2] convened in 1959 by the United Nations, is responsible for devising measures aimed at the prevention of accidents, including standards for ship design, construction, equipment, operation, and manning. According to the UN's Food and Agriculture Organisation (FAO), in 2004, the world's fishing fleet consisted of four million vessels. Of these, 1.3 million were decked vessels with enclosed areas. The rest were open vessels, of which two-thirds were traditional craft propelled by sails and oars. By contrast, all decked vessels were mechanised. Of the decked vessels, 86% were found in Asia, 7.8% in Europe, 3.8% in North and Central America, 1.3% in Africa, 0.6% in South America, and 0.4% in Oceania. Most commercial fishing boats are small, usually less than 30 m (98 ft) and up to 100 m (330 ft) for a large purse seiner or factory ship. Commercial fishing vessels can be classified by architecture, the type of fish they catch, the fishing method used, or geographical origin. The classifications are set by the FAO, who classify commercial fishing vessels by the type of gear they use.

Trawlers

Trawlers are a type of fishing vessel designed to use trawl nets to catch large volumes of fish. There are seven primary types of trawlers: the outrigger, otter trawlers, pair trawlers, side trawlers, stern trawlers, freezer trawlers, and wet fish trawlers.

- *Outrigger trawlers.* Outrigger trawlers use outriggers to tow the trawl. These are commonly used to catch shrimp. One or two otter trawls can be towed from each side. Beam trawlers, employed in the North Sea for catching flatfish, are another form of outrigger trawler. Medium-sized and high-powered vessels tow a beam trawl on each side at speeds up to eight knots.

- *Beam trawlers*. Beam trawlers use sturdy outrigger booms for towing a beam trawl, one warp on each side. Double-rig beam trawlers can tow a separate trawl on each side of the trawler. Beam trawling is used in the flatfish and shrimp fisheries in the North Sea. They are medium-sized and high-powered vessels, towing gear at speeds up to eight knots. To avoid the boat capsizing if the trawl snags on the sea floor, winch brakes can be installed, along with safety release systems in the boom stays. The engine power of bottom trawlers is also restricted to 2,000 HP (1,472 KW) for further safety.
- *Otter trawlers*. Otter trawlers deploy one or more parallel trawls kept apart horizontally using otter boards. These trawls can be towed in midwater or along the bottom.
- *Pair trawlers*. Pair trawlers are trawlers which operate together towing a single trawl. They keep the trawl open horizontally by keeping their distance when towing. Otter boards are not used. Pair trawlers operate both midwater and bottom trawls.
- *Side trawlers*. Side trawlers have the trawl set over the side with the trawl warps passing through blocks, which hang from two gallows, one forward and one aft. Until the late 1960s, side trawlers were the most familiar vessel in the North Atlantic deep-sea fisheries. They evolved over a longer period than other trawler types but are now being replaced by stern trawlers.
- *Stern trawlers*. Stern trawlers have trawls, which are deployed and retrieved from the stern. Larger stern trawlers often have a ramp, though pelagic, and small stern trawlers are often designed without a ramp. Stern trawlers are designed to operate in most weather conditions. They can work alone when midwater or bottom trawling, or the two can work together as pair trawlers.
- *Freezer trawlers*. Most trawlers operating on high sea waters are freezer trawlers. They have facilities for preserving fish by freezing, allowing them to stay at sea for extended periods of time. They are medium to enormous-sized trawlers, with the same general arrangement as stern or side trawlers.
- *Wet fish trawlers*. Wet fish trawlers are trawlers where the fish is kept in the hold in a fresh/wet condition. They must operate in areas not far distant from their landing place, and the fishing time of such vessels is limited.

Seiners

Seiners use surrounding and seine nets. This is a large group ranging from open boats as small as 10 m (33 ft) in length to ocean-going vessels. There are also specialised gears that can target demersal species. Purse seiners are highly effective at targeting aggregating pelagic species near the surface. The seiner circles the shoal with a deep curtain of netting, using bow thrusters for better manoeuverability. Then, the bottom of the net is pursed (closed) underneath the fish shoal by hauling a wire running from the vessel through rings along the bottom of the net and then back to the vessel. The most important part of the fishing operation is searching for the fish shoals and assessing their size and direction of movement. Sophisticated electronics, such as echosounders, sonar, and track plotters, may be used are used to search for and track schools, assessing their size and movement, and staying connected with the school while it is surrounded with the seine net. Crows' nests may be built on the masts for further visual support. Large vessels can have observation towers and helicopter landing decks. Helicopters and spotter planes are used for detecting fish schools. The main types of purse seiners are the American seiners, the European seiners, and the Drum seiners.

American seiners have their bridge and accommodation placed forward with the working deck aft. American seiners are most common on both coasts of North America and in other areas of Oceania. The net is stowed at the stern and is set over the stern. The power block is usually attached to a boom from a mast located behind the superstructure. American seiners use Triplerollers, which is a type of purse line winch located amidships near the hauling station, near the side where the rings are taken on board. European seiners have their bridge and accommodation located more to the after part of the vessel with the working deck amidships. European seiners are most common in waters fished by European nations. The net is stowed in a net bin at the stern and is set over the stern from this position. The pursing winch is normally positioned at the forward part of the working deck. Drum seiners have the same layout as American seiners except a drum is mounted on the stern and used instead of the power block. They are used in Canada and the USA.

Tuna purse seiners are large purse seiners, normally over 45 m (147 ft), equipped to manage large and heavy purse seines for tuna. They have the same general arrangement as the American seiner, with the bridge and accommodation placed forward. A crow's nest or tuna tower is positioned at the top of the mast, outfitted with the control and manoeuvre devices. A very heavy boom, which carries the power block, is fitted at the mast. They often carry a helicopter to search for tuna schools. On the deck are three drum purse seine winches and a power block, with other specific winches to handle the heavy boom and net. They are usually equipped with a skiff. Anchor seiners have the wheelhouse and accommodation aft and the working deck amidships, thus resembling side trawlers. The seine net is stored and shot from the stern, and they may carry a power block. Anchor seiners have the coiler and winch mounted transversally amidships. Scottish seiners are configured the same as anchor seiners. The only difference is that, whereas the anchor seiner has the coiler and winch mounted transversally amidships, the Scottish seiner has them mounted transversally in the forward part of the vessel. In Asia, the seine netter usually has the wheelhouse forward and the working deck aft, in the manner of a stern trawler.

However, in regions where the fishing effort is a labour-intensive, low-technology approach, they are often undecked and may be powered by outboards motors, or even by sail.

Line vessels

Longliners use one or more long heavy fishing lines with a series of hundreds or even thousands of baited hooks hanging from the main line by means of branch lines called 'snoods'. Hand operated longlining can be operated from boats of any size. The number of hooks and lines handled depends on the size of vessel, the number of crew, and the level of mechanisation. Large purpose-built longliners can be designed for single species fisheries such as tuna. On such larger vessels, the bridge is usually placed aft, and the gear is hauled from the bow or from the side with mechanical or hydraulic line haulers. The lines are set over the stern. Automatic or semi-automatic systems are used to bait hooks and shoot and haul lines. These systems include rail rollers, line haulers, hook separators, dehookers and hook cleaners, and storage racks or drums. To avoid incidental catches of seabirds, an outboard setting funnel is used to guide the line from the setting position on the stern down to a depth of 1 or 2 m. Small-scale longliners handle the gear by hand. The line is stored into baskets or tubs, using a hand cranked line drum. Midwater longliners are usually medium-sized vessels, which operate worldwide, purpose built to catch large pelagic species. The line hauler is usually forward starboard, where the fish are hauled through a gate in the rail. The lines are set from the stern where a baiting table and chute are located. These boats need adequate speed to reach distant fishing grounds, enough endurance for continued fishing, adequate freezing storage, suitable mechanisms for shooting and hauling longlines quickly, and proper storage for fishing gear and accessories. Freezer longliners are outfitted with freezing equipment. The holds are insulated and refrigerated. Freezer longliners are medium to large, with the same typical characteristics of other longliners. Most longliners operating on the high seas are freezer longliners. Factory longliners are equipped with a processing plant, including mechanical gutting and filleting equipment accompanied by freezing facilities, as well as fish oil, fish meal, and sometimes canning plants. These vessels have a large buffer capacity. Thus, caught fish can be stored in refrigerated sea water tanks, and peaks in the catch can also be used. Freezer longliners are large ships, working the high seas with the same typical characteristics of other large longliners. Wet fish longliners keep the caught fish in the hold in the fresh/wet condition. The fish is stored in boxes and covered with ice or stored with ice in the fish hold. The fishing time of such vessels is limited, so they operate close to the landing place.

Pole and line vessels are used to catch tuna and skipjack. The fishers stand at the railing or on special platforms and fish with poles and lines. The lines have hooks, which are baited, preferably with live bait. Caught tuna are swung on board, by two to three fishermen if the tuna is big, or with an automated swinging mechanism. The tuna usually releases themselves from the barbless hook when they hit the deck. Tanks with live bait are placed round the decks, and water spray systems are used to attract the fish. The vessels are 15–45 m (49.2–147ft) in length. On smaller vessels, fishermen fish from the main deck right around the boat. With larger vessels, there are two different deck styles: the American style and the Japanese style. On American style boats, the fisher stands on platforms arranged over the side abaft amidships and around the stern. The vessel moves ahead during fishing operation. With Japanese style boats, the fisher stands at the rail in the forepart of the vessel. The vessel drifts during fishing operations.

Trollers catch fish by towing astern one of more trolling lines. A trolling line is a fishing line with natural or artificial baited hooks trailed by a vessel near the surface or at a certain depth. Several lines can be towed at the same time using outriggers to keep the lines apart. The lines can be hauled in manually or by small winches. A length of rubber is often included in each line as a shock absorber. The trolling line is towed at a speed depending on the target species, from 2.3 knots up to at least seven knots. Trollers range from small open boats to large-refrigerated vessels 30 m (98 ft) long. In many tropical artisanal fisheries, trolling is done with sailing canoes with outriggers for stability. With intelligently designed vessels, trolling is an economical and efficient way of catching tuna, mackerel, and other pelagic fish swimming close to the surface. Purpose-built trollers are usually equipped with two or four trolling booms raised and lowered by topping lifts, held in position by adjustable stays. Electrically powered or hydraulic reels can be used to haul in the lines.

With Jiggers, there are two types of boat: specialised squid jiggers, which work mostly in the southern hemisphere and smaller vessels using jigging techniques in the northern hemisphere for catching cod. Squid jiggers have single or double drum jigger winches lined along the rails around the vessel. Strong lamps, up to 5,000 W each, are used to attract the squid. These are arranged 50–60 cm (19–23 in) apart, either as one row in the centre of the vessel, or two rows, one on each side. As the squid are caught, they are transferred by chutes to the processing plant of the vessel. The jigging motion can be produced mechanically by the shape of the drum or electronically by adjustment to the winch motor. Squid jiggers are often used during the day as midwater trawlers and during the night as jiggers. Cod jiggers use single jigger machines and do not use lights to attract the fish. The fish are attracted by the jigging motion and artificial bait (Figure 7.3).

Other vessels

Dredgers use a dredge for collecting molluscs from the seafloor. There are three types of dredges: (a) the dredge can be dragged along the seabed, scooping the shellfish from the ground. These dredges are towed in a manner

Figure 7.3 Fisheries factory ship *Kiel* at Cuxhaven, Germany.

similar to beam trawlers, and large dredgers can work three or more dredges on each side; (b) heavy mechanical dredging units are operated by special gallows from the bow of the vessel; and (c) the dredger employs a hydraulic dredge which uses a powerful water pump to operates water jets which flush the molluscs from the bottom. Dredgers do not have a typical deck arrangement. The bridge and accommodation can be aft or forward. Derricks and winches may be installed for lowering and lifting the dredge. Echosounders are used for determining depths. On inland and inshore waters, gillnets can be operated from open boats and canoes. In coastal waters, they are operated by small-decked vessels which can have their wheelhouse either aft or forward. In coastal waters, gillnetting is often used as a second fishing method by trawlers or beam trawlers, depending on fishing seasons and targeted species. For offshore fishing, or fishing on the high seas, medium-sized vessels using drifting gillnets are called drifters, and the bridge is usually located aft. The nets are set and hauled by hand on small open boats. Larger boats use hydraulic or occasionally mechanical net haulers, or net drums. These vessels can be equipped with an echosounder, although locating fish is more a matter of the fishermen's personal knowledge of the fishing grounds rather than depending on special detection equipment. Set netters also operate gillnets. However, during fishing operations, the vessel is not attached to the nets. The size of the vessels varies from open boats to large-specialised drifters operating on the high seas. The wheelhouse is usually located aft, and the front deck is used for handling gear. Normally, the nets are set at the stern by steaming ahead. Hauling is done over the side at the forepart of the deck, usually using hydraulic driven net haulers. Wet fish is packed in containers chilled with ice. Larger vessels might freeze the catch (Figure 7.4).

Figure 7.4 Typical scallop dredger, St Monans Harbour, England.

Lift netters are equipped to operate lift nets, which are held from the vessel's side and raised and lowered by means of outriggers. Lift netters range from open boats about 10 m (32 ft) long to larger vessels with open ocean capability. Decked vessels usually have the bridge amidships. Larger vessels are often equipped with winches and derricks for handling the lifting lines, as well as outriggers and light booms. They can be fitted with powerful lights to attract and aggregate the fish to the surface. Open boats are usually unmechanised or use hand-operated winches. Electronic equipment, such as fish finders and sonar and echo sounders, are used extensively on larger boats.

Trap setters are used to set pots or traps for catching fish, crabs, lobsters, crayfish, and other similar species. Trap setters range in size from open boats operating inshore to larger decked vessels, 20–50 m (65–164 ft) long, operating out to the edge of the continental shelf. Small-decked trap setters have the wheelhouse either forward or aft with the fish hold amidships. They use hydraulic or mechanical pot haulers. Larger vessels have the wheelhouse forward, and are equipped with derricks, davits, or cranes for hauling pots aboard. Locating fish is often more a matter of the fishermen's knowledge of the fishing grounds rather than the use of special detection equipment. Decked vessels are usually equipped with an echosounder, and large vessels may also have a Loran or GPS. Handliners are normally undecked vessels used for handlining (fishing with a line and hook). Handliners include canoes and other small or medium-sized vessels. Traditional handliners are less than 12 m (39 ft) long, and do not have special gear handling, there is no winch or gurdy. Locating fish is left to the fishermen's personal knowledge of fishing grounds rather than the use of special electronic equipment. Non-traditional handliners can set and haul using electrical or hydraulic powered reels. These mechanised reels are normally fastened to the gunwale or set on stanchions close to or overhanging the gunwale. They operate all over the world, some in shallow waters, some fishing up to 300 m (984 ft) deep. No typical deck arrangement exists for handliners.

Multipurpose vessels are vessels, which are designed so that they can deploy more than one type of fishing gear without major modifications to the vessels. The fish detection equipment present on board also changes according to which fishing gear is being used. Fisheries research vessels require platforms, which are capable of towing diverse types of fishing nets, collecting plankton or water samples from a range of depths, and carrying acoustic fish-finding equipment. Fisheries research vessels are often designed and built along the same lines as a large fishing vessel, but with space given over to laboratories and equipment storage, as opposed to storage of the catch. An example of a fisheries research vessel is the Scottish vessel *FRV Scotia* (Figure 7.5).

In this chapter, we have discussed some of the main types of fishing vessels used around the world. In the next chapter, we will briefly explore the role and function of research vessels.

Figure 7.5 *FRV Scotia*, entering Aberdeen Harbour, Scotland.

NOTES

1 This was originally called the Inter-Governmental Maritime Consultative Organisation (IMCO) until 1982, when it changed its name to International Maritime Organisation (IMO).
2 There is no internationally recognised 200 mile fishing limit. There is however an Exclusive Economic Zone (EEZ), which may exceed up to 200 nm in which fishing may be carried out, if a country has claim to an EEZ, as this is not an automatic right. Sometimes, when States are semi-adjacent, they may need to come to an agreement as to the extent of their EEZs. The EEZ is referred to under Articles 55 to 75 of the 1982 UN Convention on the Law of the Sea (UNCLOS).

Chapter 8

Research and scientific vessels

A research vessel (RV or R/V) is a ship or boat designed, modified, or equipped to conduct research at sea. Research vessels conduct several roles. Some of these roles can be combined into a single vessel, but others require a dedicated vessel. Due to the demanding nature of the work, research vessels are often constructed around an icebreaker hull, allowing them to operate in polar waters. The research ship has its origins in the early voyages of exploration. By the time of James Cook's *Endeavour*, the essentials of what today we would call a research ship were clearly apparent. In 1766, the Royal Society hired Cook to travel to the Pacific Ocean to observe and record the transit of Venus across the Sun. The *Endeavour* was a sturdy vessel, well designed and equipped for the ordeals she would face and fitted out with facilities for her 'research personnel', Joseph Banks. As is common with contemporary research vessels, *Endeavour* also conducted more than one kind of research, including comprehensive hydrographic survey work. Some other notable early research vessels were *HMS Beagle* (1820), *HMS Challenger* (1858), *Terra Nova* (1884), *Endurance* (1912), and *HMS* (later *RV) Calypso* (1943). Many of the names of early research vessels have been used to name later research vessels, as well as Space Shuttles. In terms of types of research and scientific ships, there are several core categories, including hydrographic survey, oceanographic survey, fisheries research, naval research, polar research, and oil and gas exploration research ships.

CATEGORIES OF RESEARCH VESSEL

There are many distinct types of research vessels employed around the world. In this chapter, we will briefly examine some of the most common and historically important types of research vessels.

Hydrographic survey

A hydrographic survey ship is a vessel designed to conduct hydrographic research and survey. Nautical charts are produced from this information

DOI: 10.1201/9781003342366-10

to ensure safe navigation by military and civilian shipping. Hydrographic survey vessels also conduct seismic surveys of the seabed and the underlying geology. Apart from producing the charts, this information is useful for detecting geological features likely to bear oil or gas. These vessels usually mount equipment on a towed structure, for example, air cannons used to generate shock waves that sound strata beneath the seabed, or mounted on the keel, for example, a depth sounder. In practice, hydrographic survey vessels are often equipped to perform multiple roles. Some also function as oceanographic research ships. Naval hydrographic survey vessels often do naval research, for example, on submarine detection. Examples of hydrographic survey vessels include *HMS Hydra* (1964) and the *CCGS Frederick G. Creed* (1988) *and HMS Enterprise (2002)* (Figure 8.1).

Oceanographic research

Oceanographic research vessels conduct research on the physical, chemical, and biological characteristics of water, the atmosphere, and climate, and to these ends carry equipment for collecting water samples from a range of depths, including the deep seas, as well as equipment for the hydrographic sounding of the seabed, along with numerous other environmental sensors. These vessels often also carry scientific divers and unmanned underwater vehicles. Since the requirements of both oceanographic and hydrographic research are vastly different from those of fisheries research, these boats

Figure 8.1 Royal Navy hydrographic vessel *HMS Enterprise*.

Figure 8.2 *NOAAS Ronald H. Brown.*

often fulfil dual roles. Recent oceanographic research campaigns include GEOTRACES and NAAMES. Examples of oceanographic research vessels include the *NOAAS Ronald H. Brown* and the Chilean Navy vessel *Cabo de Hornos* (Figure 8.2).

Autonomous research vessels

Mechanically and hydrodynamically, autonomous research platforms are like towed sonar 'fish' and ROVs. Operationally, however, known autonomous vehicles were initially limited to rivers, lakes, and bays due to navigation and collision-avoidance issues. Thus, progress in surface vessels lagged both unmanned aerial vehicles and submarine-types, due to their simpler operating conditions. Nevertheless, advances in navigation, machine vision and other sensors, and processing have spread ASVs from highly advanced models to the undergraduate level.

Technical research ships

The United States Navy used technical research ships throughout the 1960s to gather intelligence by monitoring, recording, and analysing wireless electronic communications of nations in various parts of the world. At the time, these ships were active, the mission of the ships was covert, and discussion of the true mission was prohibited ('classified information'). The mission of the ships was publicly given as conducting research into atmospheric and communications phenomena. However, the true mission was an open secret, and the ships were commonly referred to as 'spy ships'. These ships carried a crew of US Navy personnel whose specialty was intercepting wireless electronic communications and gathering intelligence from those communications [signals intelligence, communications intelligence, and electronic signals intelligence (SIGINT)]. In the 1960s, those personnel had a US Navy rating of Communications Technician (later changed to Cryptologic Technician), or CT. To transmit intelligence information that had been gathered back to United States for further processing and analysis, these ships had a special system named Technical Research Ship Special Communications, or TRSSCOM (pronounced tress-com). This Earth-Moon-Earth (EME) communications system used a special gyroscope-stabilised 4.9 m (16 ft) parabolic antenna. Radio signals were transmitted towards the moon, where they would bounce back towards the Earth and are received by a large 26 m (84 ft) parabolic antenna at a Naval Communications Station in Cheltenham, Maryland (near Washington, DC) or Wahiawa, Hawaii. Communications could occur only when the moon was visible simultaneously at the ship's location and in Cheltenham or Wahiawa. The gyro stabilisation of the antenna kept the antenna pointed at the moon while the ship rolled and pitched on the surface of the ocean. These ships were classified as naval auxiliaries with a hull designation of AGTR, which stands for Auxiliary, General, Technical Research. Five of these ships were built with hull numbers of 1–5. The first three ships of this type (*USS Oxford, USS Georgetown*, and *USS Jamestown*) were converted from World War II-era Liberty ships. The last two ships (*USS Belmont* and *USS Liberty*) were converted from Victory ships. The former Liberty ships' top speed of 11 knots (13 mph; 20 km/h) limited the first three AGTRs to missions of slow steaming on station with a minimum of transits. The sustained speed of victory ships of 18 knots (21 mph; 33 km/h) enabled *USS Belmont* to shadow Mediterranean Sea operations of the Soviet helicopter carrier *Moskva* in 1969. All the technical research ships were decommissioned and stricken by 1970. One of these ships' crew received a Presidential Unit Citation for heroism in combat. *USS Liberty* was attacked, severely damaged, and 34 crew members killed by shelling, napalm bombing, and torpedoing from Israeli jet fighter aircraft and motor torpedo boats on 8 June 1967. *USS Jamestown* was awarded a Meritorious Unit Commendation along with *USS Oxford*. The citation reads (in part)

> for meritorious service from 1 November 1965 to 30 June 1969 while participating in combat support operations in Southeast Asia. Through research and the compilation of extremely valuable technical data, *USS Jamestown* and *USS Oxford* contributed most significantly to the overall security of the United States and other Free World forces operating

in support of the Republic of Vietnam. Signed E.R. Zumwalt, Admiral, USN, Chief of Naval Operations.

Environmental Research Ships (AGER)

Three smaller ships, formerly Army Freight Supply (FS) ships converted by the US Navy to Light Cargo Ship (AKL) vessels and then to Banner-class environmental research ships, had a similar mission. In contrast to the high freeboard of the AGTR Liberty and Victory hulls, the AGER decks were low and vulnerable to boarding from a small craft. *USS Pueblo*, technically still in commission, has been held by North Korea since its attack and capture on 23 January 1968. Three technical research ships were operated as USNS ships with a Military Sea Transportation Service civilian crew and a US Navy detachment conducting the mission operations. Two ships were Maritime Commission C1-M-AV1 types. One, *USNS LT. James E. Robinson*, was a VC2-S-AP2 (Victory) type that operated in this role December 1962–April 1964 before being reclassified as AK-274 and resuming cargo operations.

Weather ships

A weather ship, or ocean station vessel, was a ship stationed in the ocean for surface and upper air meteorological observations for use in weather forecasting. They were primarily located in the north Atlantic and north Pacific oceans, reporting via radio. The vessels aided in search and rescue operations, supported transatlantic flights, functioned as research platforms for oceanographers, monitored marine pollution, and aided weather forecasting by weather forecasters and in computerised atmospheric models. Research vessels remain heavily used in oceanography, including physical oceanography and the integration of meteorological and climatological data in Earth system science. The idea of a stationary weather ship was proposed as early as 1921 by Météo-France to help support shipping and the coming of transatlantic aviation. They were used during World War II but had no means of defence, which led to the loss of several ships and many lives. Overall, the establishment of weather ships proved to be so useful during World War II for Europe and North America that the International Civil Aviation Organisation (ICAO) established a global network of weather ships in 1948, with 13 to be supplied by Canada, the United States, and some European countries. This number was eventually cut to nine. The agreement of the use of weather ships by the international community ended in 1985. Weather ship observations proved to be helpful in wind and wave studies, as commercial shipping tended to avoid weather systems for safety reasons, whereas the weather ships did not. They were also helpful in monitoring storms at sea, such as tropical cyclones. Beginning in the 1970s, their role was superseded by cheaper weather buoys. The removal of a weather ship became a negative factor in forecasts leading up to the Great Storm of 1987. The last weather ship was *Polarfront*, known as weather station *M* ('Mike'), which was removed from operation on 1 January 2010. Weather observations from ships continue from a fleet of voluntary merchant vessels in routine commercial operation (Figure 8.3).

The primary purpose of an ocean weather vessel was to take surface and upper air weather measurements and report them via radio at the synoptic hours of 0000, 0600, 1200, and 1800 Universal Coordinated Time (UTC). Weather ships also reported observations from merchant vessels, which were reported by radio back to their country of origin using a code based on the 10 mi (16 km) square in the ocean within which the ship was located. The vessels engaged in search and rescue operations involving aircraft and other ships. The vessels themselves had search radar and could activate a homing beacon to guide lost aircraft towards the ships' known locations. Each ship's homing beacon used a distinctly different frequency. In addition, the ships provided a platform where scientific and oceanographic research could be conducted. The role of aircraft supports gradually changed after 1975, as the jet aircraft began using polar routes. By 1982, the ocean weather vessel role had changed too, and the ships were used to support short range weather forecasting, in numerical weather prediction computer programmes which forecast weather conditions several days ahead, for climatological studies, marine forecasting, and oceanography, as well as monitoring

Figure 8.3 Weathership *Chofumaru 2*.

pollution out at sea. At the same time, the transmission of the weather data using Morse Code was replaced by a system using telex-over-radio.

In the 1860s, Britain began connecting coastal lightships with submarine telegraph cables so that they could be used as weather stations. There were attempts to place weather ships using submarine cables far out into the Atlantic. The first of these was in 1870 with the retired corvette *Brick* some fifty miles off Land's End, Cornwall. Although some £15,000 was spent on the project, it failed. In 1881, there was a proposal for a weather ship in the mid-Atlantic, but it came to nothing. Deep-ocean weather ships had to await the commencement of radio telegraphy. The director of France's meteorological service, Météo-France, proposed the idea of a stationary weather ship in 1921 to aid shipping and the coming of transatlantic flights. Another early proposal for weather ships occurred in connection with aviation in August 1927, when the aircraft designer Grover Loening stated that 'weather stations along the ocean coupled with the development of the seaplane to have an equally long range, would result in regular ocean flights within ten years'. During 1936 and 1937, the British Meteorological Office (Met Office) installed a meteorologist on board a North Atlantic cargo steamer to take special surface weather observations and release pilot balloons to measure the winds aloft at the synoptic hours of 0000, 0600, 1200, and 1800 UTC. In 1938 and 1939, France established a merchant ship as the first stationary weather ship, which took surface observations and launched radiosondes to measure weather conditions aloft.

Starting in 1939, United States Coast Guard vessels were being used as weather ships to protect transatlantic air commerce, as a response to the crash of Pan American World Airways *Hawaii Clipper* during a transpacific flight in 1938. The Atlantic Weather Observation Service was authorised by President Franklin Roosevelt on 25 January 1940. The Germans began to use weather ships in the summer of 1940. However, three of their four ships had been sunk by 23 November, which led to the use of fishing vessels for the German weather ship fleet. Their weather ships were out to sea for 3–5 weeks at a time and German weather observations were encrypted using Enigma machines. By February 1941, five 100 m (327 ft) United States Coast Guard cutters were used in weather patrol, usually deployed for 3 weeks at a time, and then sent back to the port for 10 days. As World War II continued, the cutters were needed for the war effort, and by August 1942, six cargo vessels had replaced them. The ships were fitted with two deck guns, anti-aircraft guns, and depth charges, but lacked SONAR (ASDIC), Radar, and HF/DF, which may have contributed to the loss of the *USCGC Muskeget* (WAG-48) with 121 personnel on board on 9 September 1942. In 1943, the United States Weather Bureau recognised their observations as 'indispensable' during the war effort. The flying of fighter planes between North America, Greenland, and Iceland led to the deployment of two more weather ships in 1943 and 1944. Great Britain established one of their own 50 mi (80 km) off the west coast of Scotland. By May 1945, frigates were used across the Pacific for similar operations. Weather Bureau personnel stationed on weather ships were asked voluntarily to accept the assignment. In addition to surface weather observations, the weather ships would launch radiosondes and release pilot balloons, or PIBALs, to determine weather conditions aloft. However, after the war ended, the ships were withdrawn from service, which led to a loss of upper air weather observations over the oceans. Due to its value, operations resumed after World War II because of an international agreement signed in September 1946, which stated that at least 13 ocean weather stations would be maintained by the US Coast Guard, with five others maintained by Great Britain and two by Brazil.

The establishment of weather ships proved to be so useful during World War II that the ICAO had established a global network of 13 weather ships by 1948, with seven operated by the United States, one operated jointly by the United States and Canada, two supplied by the United Kingdom, one maintained by France, one a joint venture by the Netherlands and Belgium, and one shared by the United Kingdom, Norway, and Sweden. The United Kingdom used Royal Navy corvettes to operate their two stations, and staffed crews of 53 Met Office personnel. The ships were out at sea for 27 days, and in port for 15 days. Their first ship was deployed on 31 July 1947. During 1949, the Weather Bureau planned to increase the number of United States Coast Guard weather ships in the Atlantic from five at the beginning of the year to eight by the year's end. Weather Bureau employees on board the vessels worked 40–63 hours per week. Weather ship *G* (George) was dropped from the network on 1 July 1949, and Navy weather ship *Bird Dog* ceased operations on 1 August 1949. In the Atlantic, weather vessel *F* (Fox) was discontinued on 3 September 1949, and there was a change in location for ships *D* (Dog) and *E* (Easy) at the same time. Navy weather ship *J* (Jig) in the north-central Pacific Ocean was taken out of service on 1 October 1949. The original international agreement for a 13-ship minimum was later amended downward. In 1949, the minimum number of weather ships operated by the United States was decreased to 10, and in 1954, the figure was lowered again to nine, both changes being made for economic reasons. Weather vessel *O* (Oboe) entered the Pacific portion of the network on 19 December 1949. Also in the Pacific, weather ship *A* (Able) was renamed ship P (Peter) and moved 200 mi (320 km) to the east-northeast in December 1949, while weather vessel *F* (Fox) was renamed *N* (Nan).

Weather ship *B* (Baker), which had been jointly operated by Canada and the United States, became solely a United States venture on 1 July 1950. The Netherlands and the United States began to jointly operate weather ship *A* (Able) in the Atlantic on 22 July 1950. The Korean War led to the discontinuing of weather vessel *O* (Oboe) on 31 July 1950 in the Pacific, and ship *S* (Sugar) was established on 10 September 1950. Weather ship *P*'s (Peter) operations were taken over by Canada on 1 December 1950, which allowed the Coast Guard to begin operating station U (Uncle) 1,200 mi

(2,000 km) west of northern Baja California on 12 December 1950. As a result of these changes, ship *N* (Nan) was moved 250 mi (400 km) to the southeast on 10 December 1950. Responsibility for weather ship *V* (Victor) transferred from the US Navy to the US Coast Guard and Weather Bureau on 30 September 1951. On 20 March 1952, vessels *N* (November) and *U* (Uncle) were moved 20–30 mi (32–48 km) to the south to lie under airplane paths between the western United States coast and Honolulu, Hawaii. In 1956, *USCGC Pontchartrain*, while stationed with *N* (November), notably rescued the crew and passengers of Pan Am Flight 6 after the crippled aircraft diverted to the clipper's position and ditched in the ocean. Weather vessel *Q* (Quebec) began operation in the north-central Pacific on 6 April 1952, while in the western Atlantic, the British corvettes used as weather ships were replaced by newer *Castle-class* frigates between 1958 and 1961. In 1963, the entire fleet won the Flight Safety Foundation award for their distinguished service to aviation. In 1965, there were a total of 21 vessels in the weather ship network. Nine were from the United States, four from the United Kingdom, three from France, two from the Netherlands, two from Norway, and one from Canada. In addition to the routine hourly weather observations and upper air flights four times a day, two Soviet ships in the northern and central Pacific Ocean sent meteorological rockets up to a height of 50 mi (80 km). For a time, there was a Dutch weather ship stationed in the Indian Ocean. The network left the Southern Hemisphere uncovered. South Africa maintained a weather ship near latitude 40 degrees north, longitude 10 degrees east between September 1969 and March 1974.

When compared with the cost of unmanned weather buoys, weather ships were prohibitively expensive, with weather buoys quickly replacing the American weather ships throughout the 1970s. Across the northern Atlantic, the number of weather ships dwindled over the years. The original nine ships in the region had fallen to eight after ocean vessel *C* (Charlie) was discontinued by the USA in December 1973. In 1974, the US Coast Guard announced plans to terminate all United States stations, and the last American weather ship was replaced by a newly developed weather buoy in 1977. A new international agreement for ocean weather vessels was reached through the World Meteorological Organisation (WMO) in 1975, which eliminated Ships *I* (India) and *J* (Juliet), and left ships *M* (Mike), *R* (Romeo), *C* (Charlie), and *L* (Lima) across the northern Atlantic, with the four remaining ships in operation through 1983. Two of the British frigates were refurbished, as there was no funding available for new weather ships. Their other two ships were retired, as one of the British run stations was eliminated in the international agreement. In July 1975, the Soviet Union began to maintain weather ship *C* (Charlie), which it would operate through the remainder of the 1970s and 1980s. The last two British frigates were retired from ocean weather service on 11 January 1982, but the international agreement for weather ships was continued through to 1985. Because of high operating costs and budget issues, weather ship *R* (Romeo) was recalled from the Bay of Biscay before the deployment of a weather buoy for the region. This recall was blamed for the minimal warning given in advance of the Great Storm of 1987, when wind speeds of up to 93mph (149 km/h) caused extensive damage to the south of England and the north of France. The last weather ship to operate was *Polarfront*, known as weather station *M* (Mike) at 66 degrees north, 02 degrees east, run by the Norwegian Meteorological Institute. *Polarfront* was withdrawn from operation on 1 January 2010. Despite the loss of designated weather ships, weather observations from ships continue from a fleet of voluntary merchant vessels in routine commercial operation, whose number has consistently decreased since 1985.

Beginning in 1951, British ocean weather vessels began oceanographic research, such as monitoring plankton, casting of drift bottles, and sampling seawater. In July 1952, as part of a research project on birds by Cambridge University, 20 shearwaters were taken more than 100 mi (161 km) offshore on British weather ships, before being released to see how quickly they would return to their nests, which were more than 450 mi (720 km) away on Skokholm Island [2.5 mi (4 km) off the coast of Pembrokeshire, Wales]. Eighteen of the 20 birds returned, the first in just 36 hours. During 1954, British weather ocean vessels began to measure sea surface temperature gradients and ocean waves. In 1960, the weather ships proved to be helpful in ship design through a series of recordings made on paper tape which evaluated wave height, pitch, and roll. They were also useful in wind and wave studies, as they did not avoid weather systems, as merchant ships tended to, and were considered a valuable resource. In 1962, British weather vessels measured sea temperature and salinity values from the surface down to 3,000 m (9,800 ft) as part of their studies. Upper air soundings launched from weather ship *E* (Echo) were of great utility in determining the cyclone phase of Hurricane Dorothy in 1966. During 1971, British weather ships sampled the upper 500 m (1,600 ft) of the ocean to investigate plankton distribution by depth. In 1972, the Joint Air-Sea Interaction Experiment (JASIN) used special observations from weather ships for their research. More recently, in support of climate research, 20 years of data from the ocean vessel *P* (Papa) was compared with nearby voluntary weather observations from mobile ships within the International Comprehensive Ocean-Atmosphere Data Set to check for biases in mobile ship observations over that period.

Tropical meteorology

Between the years 1959 and 1965, 40 research ships from 23 countries assayed the Indian Ocean, which till then, was the most unexplored maritime environment on earth. In 1958, the American ship *Vema* began the first purposeful assay of the ocean, with additional ships joining between 1959 and 1960: *Diamantina* (Australia), *Commandant Robert Giraud* (France), *Vitjaz* (USSR*), Argo, Requisite* (USA), and the *Umitaka Maru* (Japan). West Germany followed with the Meteor in 1964/1965. With the commencement of the Global Atmospheric Research Programme (GARP) in 1967, the WMO and the International Council of Scientific Unions began a global weather experiment in which the entire atmosphere and the ocean surface would be observed for the first time. More than 50 ships worked in the equatorial areas around the globe, collecting oceanographic and meteorological measurement data for an 'inventory of world weather'.

Fisheries research

A fisheries research vessel requires platforms capable of towing several types of fishing nets, collecting plankton or water samples from a range of depths, and carrying acoustic fish-finding equipment. Fisheries research vessels are often designed and built along the same lines as a large fishing vessel, but with space given over to laboratories and equipment storage, as opposed to storage of the catch. Examples of fisheries research vessel include CEFAS' *RV CEFAS Endeavour*, Marine Scotland's *FRV Scotia* (2009), and the Danish fisheries research vessel *Dana IV* (1980) (Figure 8.4).

Figure 8.4 RV CEFAS Endeavour at Lowestoft, England.

Naval research

Naval research vessels investigate naval concerns, such as submarine and mine detection or sonar and weapons trials. An example of a naval research vessel is the German Navy's 3,500 tonne *Planet*, which is the most modern naval research vessel within NATO.

Polar research

Polar research vessels are constructed around an icebreaker hull, allowing them to engage in ice navigation and operate in polar waters. These vessels usually have dual roles, particularly in the Antarctic, where they function also as polar replenishment and supply vessels to the Antarctic research bases. Examples of polar research vessels include the *USCGC Polar Star* (1976), *RSV Aurora Australis* (1989), *RSV Nuyina* (2015), and the *RRS David Attenborough* (2018) (Figure 8.5).

Oil exploration

Oil exploration is performed in several ways, one of the most common being mobile drilling platforms or ships that are moved from area to area as needed to drill into the seabed to find out what deposits lie beneath it (Figure 8.6).

Figure 8.5 RRS David Attenborough.

Figure 8.6 Petroleum Geo-Services owned seismic vessel *Ramform Challenger*, Valletta Harbour, Malta.

DEVELOPMENT OF THE RESEARCH VESSEL

Race to the poles (19th century)

At the end of the 19th century, there was intense international interest in exploring the North and South Poles. The search operations for the lost Franklin expedition were barely forgotten as Russia, Great Britain, Germany, and Sweden set new scientific tasks for the Arctic Ocean. In 1868, the Swedish ship *Sofia* conducted temperature measurements and oceanographic observation in the sea area around Svalbard. During this year, the vessel *Greenland*, built in Norway, operated in the same area under the German command of Carl Koldeway. In 1868–1869, the ship owner A. Rosenthal gave scientists the opportunity to come on board his whaling trips and by 1869, the ship *Germania*, which was escorted by the *Hansa* and led the Second German North Polar Expedition, was built. The *Germania* returned safely from the expedition and was used later for further research. The *Hansa*, in contrast, was crushed by ice and sunk. In 1874, the Austrian-Hungarian *Tegetthoff* as well as the American schooner Polaris met with the same fate. The Royal Navy ships *HMS Alert* and *HMS Discovery* of the British Arctic Expedition of 1875–1876 were more successful. In 1875, they left Portsmouth to cross the Arctic Ocean and reach as close as possible to the pole. Although they did not reach the pole itself, they brought plenty of precious observations back. During these years, Adolf Erik Nordenskiöld's discovery of the Northeast Passage and first circumnavigation of Eurasia during the Vega expedition outstood all other expeditions. Fridtjof Nansen's famous *Fram* took deep-sea soundings and conducted hydrographical, meteorological, and magnetic surveys throughout the polar basin. In 1884, a press item stating that the American *Jeanette* had sunk near the Siberian coast three years earlier, inspired Fridtjof Nansen (Figure 8.7).

Three years following the loss of the *Jeanette*, some items of equipment including sou'wester pants were found washed up on the southwestern Greenland coast. When Nansen heard about this, he deduced that there was only explanation. These pieces had drifted surrounded by ice via the polar basin and along the east coast of Greenland until they ended up in Julianehab. Bits of driftwood, used by the Inuit, were even more enlightening for Nansen, as they could only have derived from the areas of the Siberian rivers that flowed into the Arctic Ocean. Nansen realised there was a current which floated from somewhere between the pole and Franz Josef Land through the Arctic Ocean to the east coast of Greenland. Nansen planned to freeze a ship into the ice and to let it float. Unlike other explorers, Nansen supposed that a well-constructed ship would take him to the North Pole safely. The success of the expedition depended on the ship's construction, especially its resistance against ice pressure. Indeed, neither the *Fram* nor Nansen, who impatiently left his ship and ventured towards the pole on foot, achieved

Figure 8.7 Discovery, Antarctica, circa 1901.

their goal but Nansen's theory about the currents was proved correct. While Nansen was returning from the ice of the Arctic Ocean, his compatriot Roald Amundsen set off to the Antarctic. On board, the *Belgica* Amundsen accompanied the Belgian Antarctic Expedition as a steersman. For this expedition, Adrien de Gerlache purchased the whale catcher *Patria* for 70,000 francs, overhauled the engine, installed additional cabins, arranged a laboratory, and renamed the ship *Belgica*. Between 1887 and 1899, biological and physical observations were conducted to the west of the Antarctic Peninsula and to the south of Peter I Island. The *Belgica* became the first ship to remain over winter in the Antarctic. In 1895, Georg Neumayer, director of the Hamburg Naval Observatory, launched the slogan 'off to the South Pole' at the Sixth International Geographic Congress. Motivated by reports of the Norwegian Carsten Borchgrevink, who was the first person to land on the new continent, the Congress declared the exploration of Antarctica as the most urgent task for the next several years.

20th century

While Borchgrevink, steering towards the Southern Cross, was on target for Antarctica, a German, Swedish, and British expedition was preparing to explore the Southern Ocean. Germany built the expedition ship *Gauss* for 1.5 million marks at the Howaldtswerke in Kiel. Designed off the model of the *Fram*, the *Gauss*, which weighed 1,442 tonnes and was 46 m (151 ft) long, had a round hulk to withstand the ice pressure. The *Gauss* had three masts and one auxiliary engine of 275 horsepower (205 kW). With a 60-strong crew, she could operate for almost three years without needing assistance. From 1901 until 1903, Erich von Drygalski led the German Antarctic expedition and conducted extensive studies in the southern part of the Indian Ocean. The Swedish expedition under the command of Otto Nordenskjöld used the old *Antarctic*, weighing only 353 tonnes, which had already been used by Borchgrevink in 1895. The expedition, intending to spend the winter in the Antarctic Peninsula, was ill-fated from the beginning. In 1902, the *Antarctic* sank. Fortunately, the Argentine gunboat *Uruguay* rescued all crewmembers. At the same time, the British Government prompted Dundee Shipbuilders in Scotland to build a ship for Robert Falcon Scott's expedition. The *Discovery*, weighing 1,620 tonnes, and 52 m (171 ft) in length, was fitted with an auxiliary motor of 450 horsepower. Nevertheless, during the expedition, the ship became frozen by ice. Only the relief ship *Morning*, sent by the British Admiralty to free her, was able to free her and with *Terra Nova* escorted the *Discovery* back home. Undeterred, the British decided to return to the Antarctic, this time under the command of the Scottish naturalist, William Spiers Bruce. Bruce had collaborated with the whale catchers *Balaen*a and *Activ*e in the Southern Ocean throughout 1892 and 1893. He hoped that he could acquire the field-tested *Balaena* but found the ship too expensive, buying instead the Norwegian whale catcher *Hekla* for £2620, a ship that had sailed under the Danish flag along Greenland's coast in 1891 and 1892. For another £8000, he had the ship repaired and provided it with a new engine. Under the new name *Scotia*, the ship completed its way into the Southern Ocean.

At that point, the French vessel *Français*, a 32 m (105 ft) three-master arrived in the Antarctic with the French doctor and naturalist Jean-Baptiste Charcot on board. After his return from the Southern Ocean, Charcot sold the *Français*, but in 1908 returned to Antarctica, this time on board the *Pourquoi Pas*.

Sir Ernest Shackleton, a member of the '*Discovery-Expedition*' of 1901, returned to the Antarctic with the forty-year-old *Nimrod*. His plan was to march to the South Pole but was forced to give up just 111 mi (180 km) before reaching his goal. Shackleton had at first tried to purchase the whale catcher *Bjørn* built at the Risør shipyard in Lindstøl, Norway. As Shackleton could not raise sufficient financing for the expedition, he conceded the ship to Wilhelm Filchner's second German Antarctic Expedition. Filchner refurbished the ship at the Blohm & Voss dockyard and renamed it *Deutschland*. There was sufficient room for 34 crewmembers, while single cabins were available for the scientists and ship's officers. In addition, a geologist, meteorologist, an oceanographer, and a zoologist shared a laboratory. The expedition, which visited the Southern Ocean between 1911 and 1912, made substantial contributions to the physical and chemical conditions in the western part of the southern Atlantic Ocean and the Weddell Sea.

The Japanese arrived in the Southern Ocean in 1911. Ensign Nobu Shirase explored the eastern part of the Ross Ice Shelf with the *Kainan Maru*. The years between 1910 until 1912 were characterised by the 'great race' between the Norwegian Amundsen and the Englishman Scott. While Scott travelled in the 26-year-old *Terra Nova*, which had escorted him out of the ice in 1903, Roald Amundsen borrowed the dependable *Fram* for his South Pole expedition. During the dramatic race, Australia unobtrusively sent its first expedition ship, the *Aurora*, into the Antarctic under the leadership of Douglas Mawson. An air-tractor, the first airplane to be despatched to the region, was on board but proved to be useless. The failed attempt to cross Antarctica, for which Shackleton used the *Endurance* and Mawson's *Aurora*, was one of the last Antarctic Expeditions before the outbreak of the First World War.

Between the world wars

After the war, Shackleton was one of the first to reengage in polar exploration. For a new Arctic Expedition, he bought the *Foca I* which was designed in Norway and specified for polar areas. Organisational difficulties were encountered, and Shackleton needed to change his plans, and set course instead for the Southern Ocean. Unfortunately, Shackleton failed to complete the expedition, having died on board the *Quest* in 1922. His long-time companion Frank Wild assumed the leadership and advanced as far as the South Sandwich Islands until pack ice forced Wild to turn round and make for home. Later, the *Quest* resumed its original role as a sealer. In the period from 1930 to 1931, H. G. Watkins deployed the *Quest* for the British Air Route Expedition, surveying the eastern coast of Greenland in search of a site for an air base. After the First World War interrupted oceanographic research, international scientific activities were renewed in 1920. The invention of the echo sounder in 1912 provided new significance for international marine research. Henceforth, it was possible to measure the distance from the surface to the seabed by sending acoustic signals instead of using wires and weights. The technology was evaluated extensively throughout the war. In 1922, the American destroyer *Stewart* took the first echo profile over the North Atlantic and one year later, completed sonic logging between San Francisco and San Diego. Between 1929 and 1934, the *USS Ramapo* completed about thirty profiles of the northern Pacific Ocean. In 1927, the German cruiser *Emden* conducted a series of soundings of the ocean trench to the east of the Philippines.

The German ship *Meteor* was the first to use the echo sounder for scientific purposes in the 1920s on the German Meteor Expedition. For the first time, an ocean, the Atlantic, was systematically mapped. The *Meteor* crossed the South Atlantic from the ice line to 20 degrees north on 14 mapping passages. With 67,000 echo soundings, cartographers were able to produce a modern depth chart. Other geomagnetic and oceanographic mapping expeditions followed starting with the American research ship *Carnegie* in the Pacific Ocean from 1928 to 1929; detailed reconnaissance in Indonesia by the Dutchman Willebrord Snellius; the exploration of the waters around the Antarctic by the British *William Scoresby* and *Discovery II* and the expedition of the American schooner *Atlantis*, which sailed from the West Atlantic to the Gulf of Mexico between 1932 and 1938. The Scandinavian countries also continued their activities at the end of the 1920s. Danish scientists devoted themselves to research in marine biology. The oceanographic *Dana* Expedition, led by Johannes Schmidt and financed by the Carlsberg Foundation, was the most important Danish marine expedition of that time. In March 1914, the Canadian-born British oceanographer John Murray died, leaving in his will instructions that a new expedition be organised as soon as the necessary capital could be accumulated. In 1931, his son, John Challenger Murray set about putting his father's legacy into action. Murray had originally intended to enlist the research ships *William Scoresby, Dana*, and *George Bligh*. As all three vessels were unavailable, Murray accepted an offer from the Egyptian Government to take the *Mabahiss* instead. The *Mabaniss* departed Alexandria on 3 September 1933, returning on 25 May 1934. During this period, under the leadership of Seymour Sewell, the *Mabahiss* surveyed the Red Sea, Bay of Biscay, Indian Ocean, and the Gulf of Oman; an undertaking of more than 22,000 nautical miles (25,476 mi; 41,000 km). During the expedition, various chemical, physical, and biological assays were collected.

Eight years later, a new phase in marine research began. Unlike previously when expeditions were competitive, countries were now working together on one combined expedition. The German *Altair* and the Norwegian *Armauer Hansen* performed common measurements on the International Gulf Stream Expedition, which shed light on the fluctuation of the Gulf Stream. For this experiment, the German *Meteor*, the Danish *Dana*, and the French air-base vessel *Carimare* also collaborated to collect and deliver data. Two flying boats, the *Boreas* and the *Passat*, equipped with aerial photography equipment were positioned on board the German research ship *Schwabenland*. The application of this technology over Antarctica proved revolutionary, with stereo photography used for the first time. On the way back to Europe, the aircraft carrier conducted oceanographic, biological, and meteorological observations and every 15–30 minutes took echo soundings.

The post-war period

Following a hiatus between 1939 and 1946, the major seafaring nations once again turned their attention to marine research. Leading this was the United States although other expeditions were launched as well. The Swedish *Albatross* expedition of 1947–1948 crossed the equator no less than 18 times and covered 45,000 nautical miles (51,573 mi; 83,000 km). Over 200 cores, fastened on a perpendicular, were dragged 0.93 mi (1.5 km) over the seabed. The second Danish *Galathea* expedition, from 1950 to 1951, headed by Anton Bruun, concentrated on studying deep sea life and succeeded in catching live animals from the bottom of the Philippine Trench at a depth of 10,190 m (33,430 ft), therein proving that life can exist in even the deepest parts of the ocean. Another significant find of the expedition was the discovery of a 'living fossil', the *monoplacophoran Neopilina galathea*, dredged from the bottom of the Mexican Gulf. Already by the 1930s, American scientists had begun to use seismic measuring methods in flat waters and during the war, physicist Maurice Ewing conducted the first seismic refractions. The introduction of this geophysical method, as well as palaeomagnetic research (studying the Earth's magnetic field preserved in magnetic iron-bearing minerals), led to a reinvigoration of Alfred Wegener's continental drift theory and subsequent development of plate tectonics. Following the end of the Second World War, the number of missions escalated worldwide. For the period 1950–1960, no less than 110 expeditions were performed in the Mediterranean Sea alone.

Increasing collaboration

At the beginning of the 1950s, an innovative approach to marine research methods was emerging. In line with the NOPAC enterprise, the Canadian and American ships *Hugh M. Smith, Brown Bear, Ste. Teherse, Horizon,*

Black Douglas, Stranger, and the *Spencer F. Baird, and the Japanese vessels Oshoro Maru, Tenyo Maru, Kagoshima Maru, Satsuma, Umitaka Maru* took part in this study. Working on a grand scale, common international research programmes got underway during the international geophysical year (IGY) 1957/1958. For the exploration of deep-sea circulation and strong sea currents, 60 research ships from 40 nations were deployed, including the research ships *Crawford, Atlantis, Discovery II, Chain*, and the Argentine sounding vessel *Capitan Canepa*. During this study, 15 profiles at intervals of eight latitudes between 48 degrees north and 48 degrees south were taken. Furthermore, 20 ships from 12 different nations participated in the Cold Wall enterprise, which was another programme of the IGY. The research was focused on the Cold Wall, which separates the warm, high-salt Gulf Stream and the cold, low-salt water masses of the Arctic polar region and stretches northwest from the Newfoundland banks to the Norwegian Sea. Later, West Germany participated in the programme with the *Antorn Dohrn* and the *Gauss*. This was Germany's first participation after World War II. For the first time, the Soviet Union, which had a fleet of 20 research ships and one research dive station, took part in an international project during the IGY. Moreover, the USSR commissioned the research ships *Vityaz* (5700 tonnes) and *Michail Lomonosov* (6,000 tonnes), in addition to the icebreaker *Ob*. The IGY marked the beginning of a new exploration phase in marine research characterised by international cooperation and synoptic research.

With the Overflow programme, scientists tried to reconnoitre the overflow of the cold Arctic ground water over the submarine ridge between Iceland and the Faroe Islands. Nine research ships from five European countries participated in the study. This programme was later enlarged with thirteen research ships from seven countries soon joining the project. Denmark appointed the *Dana* and *Jens Christian Svabo*, Iceland the *Bjarni Sæmundson*, Canada the *Hudson*, Norway the *Helland Hansen*, the USSR the *Boris Davydov* and *Professor Zubov*, Great Britain the *Challenger, Shackleton* and *Explorer*, and West Germany the *Meteor*, *Walther Herwig*, and *Meerkatze II*.

This completes Part I. In Part II, we will begin to explore the design, role, and function of wet cargo ships, starting with chemical tankers.

Part II

Wet cargo ships

Chapter 9

Chemical tankers

Chemical tankers are a class of tanker ships designed to transport chemicals in bulk. As defined in MARPOL Annex II, chemical tanker refers to any ship constructed or adapted for carrying in bulk any liquid product listed in Chapter 17 of the International Bulk Chemical *(IBC)* Code. As well as industrial chemicals and clean petroleum products, these ships also often carry other types of sensitive cargo, which require a high standard of tank cleaning, such as palm oil, vegetable oils, tallow, caustic soda, and methanol. Oceangoing chemical tankers range from 5,000 metric tonnes dwt to 35,000 metric tonnes dwt in size, which is smaller than the average size of other tanker types. This is due to the specialised nature of their cargo and the size restrictions of the port terminals where they call to load and discharge. Chemical tankers normally have a series of separate cargo tanks, which are either coated with specialised coatings (such as phenolic epoxy or zinc paint) or are made from stainless steel. The coating or cargo tank material determines what types of cargo a particular tank can carry. For instance, stainless steel tanks are required for aggressive acid cargoes such as sulphuric acid and phosphoric acid, while 'easier' cargoes – such as vegetable oil – can be carried in epoxy-coated tanks. The coating or tank material also influences how quickly the ship's tanks can be cleaned. Typically, ships with stainless steel tanks can carry a wider range of cargoes and can be cleaned more quickly between one cargo and another, which justifies the additional cost of their construction.

In general, ships carrying chemicals in bulk are classed into one of the following three types:

- *Type 1.* Tankers intended to transport products with profoundly serious environmental and safety hazards requiring maximum preventive measures to prevent any leakage of cargo.
- *Type 2.* Tankers intended to transport products with appreciably severe environmental and safety hazards requiring significant preventive measures to preclude an escape of such cargo.
- *Type 3.* Tankers intended to transport products with sufficiently severe environmental and safety hazards to require a moderate degree of containment to increase survival capability in a damaged condition.

DOI: 10.1201/9781003342366-12

Figure 9.1 Typical chemical tanker, *TripleA*, outbound from Rotterdam, Netherlands.

Most chemical tankers are IMO 2 and IMO 3 rated since the volume of IMO 1 cargoes is extremely limited. Chemical tankers often have a system for tank heating to maintain the viscosity of certain cargoes, typically through passing pressurised steam through stainless steel 'heating coils' in the cargo tanks. These coils circulate fluid in the tank by convection, transferring heat to the cargo. All modern chemical tankers feature a double-hull construction. Most modern vessels have one hydraulically driven and submerged cargo pump for each tank with independent piping, which means that each tank can load separate cargo without them being mixed. Consequently, many oceangoing chemical tankers may carry numerous different grades of cargo on a single voyage, often loading and discharging these 'parcels' at different ports or terminals. This means that the scheduling, stowage planning, and operation of such ships requires an elevated level of coordination and specialist knowledge, both at sea and on shore. Tank cleaning after discharging cargo is an especially important aspect of chemical tanker operations. This is because cargo residue can adversely affect the purity of the next cargo to be loaded. Before tanks are cleaned, they must be properly ventilated and checked to be free of potentially toxic or explosive vapours. Chemical tankers usually have transverse stiffeners on deck rather than inside the cargo tanks. This ensures that the tank walls are smooth and thus easier to clean using permanently fitted tank cleaning machines. Cargo tanks, either empty or filled, are normally protected against explosion by inert gas blankets. Nitrogen is most often used as the inert gas of choice, supplied either from portable gas bottles or from a nitrogen generator.

Currently, the latest chemical tankers are built by shipbuilders in Japan, Korea, and China, where most stainless-steel chemical tankers are built. This is because welding stainless steel to the accuracy required for cargo tank construction is a difficult skill to acquire. Nations with minor builders include Turkey, Italy, Germany, and Poland. The largest chemical tanker operators include Stolt-Nielsen (UK), Odfjell (Norway), Navig8 (UK), and Mitsui O.S.K. Lines (MOL) (Japan). Charterers or the end users of the ships include oil majors, industrial consumers, commodity traders, and specialist chemical companies.

Chapter 10

FPSO and FLNG units

Floating production storage and offloading (FPSO) units are a type of floating vessel used by the offshore oil and gas industry for the production and processing of hydrocarbons, and for the storage of oil. An FPSO vessel is designed to receive hydrocarbons produced by itself or from nearby platforms or subsea templates, process them, and store the oil on board until it can be offloaded onto a tanker or, less frequently, transported through a pipeline to an onshore facility. FPSOs are preferred in frontier offshore regions as they are easy to install and do not require local pipeline infrastructure to export the oil. FPSOs can be either converted oil tankers (such as the *Knock Nevis*, ex-*Seawise Giant*, which for many years was the world's largest ship. It was converted into an FSO for offshore use before being scrapped in 2010) or a vessel built specially for the application. A vessel used only to store oil (without processing it) is referred to as a floating storage and offloading (FSO) vessel. Most FSOs are converted single hull super tankers. Recent developments in the liquefied natural gas (LNG) industry have led to the development and location of LNG processing trains or floating LNG units (FLNGs). These are typically located above remote smaller gas fields that would otherwise be uneconomical to develop. The added benefits of using mobile LNG processing units include reduced capital expenditure and a minimal impact on the marine environment. Unlike FPSOs, FLNGs allow full-scale deep processing to the same extent as onshore LNG plants but with a 25% smaller footprint. The first three purpose-built FLNGs constructed were *Prelude FLNG* by Shell, and the *PFLNG1* and *PFLNG2* for Petronas (Malaysia), all in 2016. At the other end of the LNG logistics chain, where the natural gas is brought back to ambient temperature and pressure, a specially modified ship may also be used as a floating storage and regasification unit (FSRU). LNG FSRUs receive the LNG from offloading LNG carriers, and the onboard regasification system provides natural gas exported to shore through risers and pipelines (Figure 10.1).

The oil produced from offshore production platforms can be transported to the mainland either by pipeline or by tanker. When a tanker is chosen to transport the oil, it is necessary to accumulate oil in some form of storage tank, such that the oil tanker is not continuously occupied during oil

DOI: 10.1201/9781003342366-13

Figure 10.1 FPSO Kizomba.

production and is only needed once sufficient oil has been produced to fill the tanker. Floating production, storage, and offloading vessels are particularly effective at storing oil in remote or deep-water locations, where seabed pipelines are not cost-effective. FPSOs eliminate the need to lay expensive long-distance pipelines from the processing facility to an onshore terminal. This can provide an economically attractive solution for smaller oil fields, which can be exhausted in a few years and do not justify the expense of installing a pipeline. Once the field is depleted, the FPSO is moved to a new location. New build FPSOs have a high initial cost (up to US$1 billion) but require limited maintenance. In addition, the ability to reposition and or repurpose them means that they can outlast the life of the production facility by decades. Converting an oil tanker or a similar vessel into an FPSO can cost as little as US$100 million.

DEVELOPMENT OF THE FPSO

Oil has been produced from offshore locations since the late 1940s. Originally, all oil platforms sat on the seabed, but as exploration moved to deeper waters and more distant locations in the 1970s, floating production systems were developed. The first oil FPSO was built in 1977 and installed above the *Shell Castellon* field, located in the Spanish Mediterranean. Today, over 270 vessels are deployed worldwide as oil FPSOs. In 2009, Shell and

Samsung announced an agreement to build up to 10 LNG FPSOs. In 2011, Shell announced the development of the 488 m (1,601 ft) long and 74 m (242 ft) wide floating LNG facility (FLNG), *Prelude*, which was to be situated 124 mi (200 km) off the coast of Western Australia. As of 2022, the *Prelude* is the largest and most expensive merchant vessel ever built with an estimated price tag of over US$12billion. In 2012, the Malaysian oil company Petronas signed an agreement with Technip of France and DSME of South Korea to design, construct, install, and commission the world's first floating natural gas liquefaction unit. With a length overall of 300 m (984 ft) and a beam of 60 m (196 ft), and moored 112 mi (180 km) offshore Bintulu, the *Petronas Floating LNG 1* can produce 1.2 metric-tonnes per annum (mpta) of LNG, increasing Malaysia's total gas production from 25.7 to 26.9 mpta. At the opposite (discharge and regasification) end of the LNG chain, the first ever conversion of an LNG carrier, the *Golar LNG*, into an LNG FSRU was conducted in 2007 by Keppel shipyard in Singapore (Figure 10.2).

FPSO records

The FPSO operating in the deepest waters is the *FPSO BW Pioneer*, built and operated by BW Offshore on behalf of Petrobras Americas Inc. The FPSO is moored at a depth of 2600 m (8530 ft) in Block 249 of the *Walker Ridge* field in the Gulf of Mexico and is rated for 80,000 bbl/d (13,000 m³/d). The EPCI contract was awarded in October 2007 with production starting in early 2012. The FPSO conversion was conducted at the MMHE Shipyard Pasir Gudang in Malaysia, while the topsides were fabricated in modules at

Figure 10.2 FPSO Firenze.

various international vendor locations. The FPSO has a disconnectable turret. This means that the vessel can disconnect on notification of hurricanes and reconnect with a minimal down time. A contract for an FPSO to operate in even deeper waters (2,900 m, 9,514 ft) for Shell's *Stones* field in the Gulf of Mexico was awarded to SBM Offshore in July 2013. One of the world's largest FPSO is the *Kizomba A*, with a storage capacity of 2.2 million barrels (350,000 m^3). Built at a cost of over US$800 million by Hyundai Heavy Industries in Ulsan, Korea, the *Kizomba A* is operated by Esso Exploration Angola, which is a subsidiary of the ExxonMobil conglomerate. Located in 1,200 m (3,940 ft) of water, some 200 mi (320 km) offshore from Angola, it weighs 81,000 metric-tonnes and has a length overall of 285 m (935 ft), a beam if 63 m (206 ft), and a height above the waterline of 32 m (104.9 ft).

The first FSO to operate in the Gulf of Mexico, the *FSO Ta'Kuntah*, has been in operation since August 1998. The FSO, owned and operated by MODEC, is under a service agreement with PEMEX Exploration and Production (Mexico). The vessel was installed as part of the *Cantarell* Field Development. The field is in the Bay of Campeche, off the coast of Mexico's Yucatán Peninsula. She is a converted ULCC tanker with a SOFEC external turret mooring system, two flexible risers connected in a lazy-S configuration between the turret and a pipeline end manifold (PLEM) on the seabed, and an offloading system that allows up to two tankers at a time to moor and load, in tandem or side by side. The FSO is designed to manage 800,000 bbl/d (130,000 m^3/d) with no allowance for downtime. The *Skarv FPSO*, developed and engineered by Aker Solutions for BP Norge, is one of the most advanced and largest FPSO deployed in the Norwegian Sea, positioned offshore mid-Norway. Skarv is a gas condensate and oil field development. The development ties in five sub-sea templates, and the FPSO has the capacity to include several nearby smaller wells. The process plant on the vessel can oversee approximately 19,000,000 cubic metres per day (670,000,000 cu ft/d) of gas and 13,500 cubic metres per day (480,000cu ft/d) of oil. A 49 mi (80 km) gas export pipe connects to the Åsgard transport system. Aker Solutions (formerly Aker Kvaerner) developed the front-end design for the floating production facility as well as the overall system design for the field and preparation for procurement and project management of the total field development. The hull is of an Aker Solutions proprietary 'Tentech975' design. BP also selected Aker Solutions to perform the detail engineering, procurement, and construction management assistance (EPcma) for the Skarv field development. The EPcma contract covered detail engineering and procurement work for the FPSO topsides as well as construction management assistance to BP including hull and topside facilities. The production started in the field in August 2011. BP awarded the contract for fabrication of the *Skarv FPSO* hull to Samsung Heavy Industries (South Korea) and the Turret contract to SBM. The FPSO has a length overall of 292 m (958 ft), a beam of 50.6 m (166 ft), a draught of 29 m (95 ft), and can accommodate as many as 100 people in single berth cabins.

Chapter 11

Gas carriers

A gas carrier, gas tanker, LPG carrier, or LPG tanker is a ship designed to transport LPG, LNG, CNG, or liquefied chemical gases in bulk. There are six main types of gas carriers in operation today, namely the fully pressurised gas carrier; the semi-pressurised gas carrier; ethylene and gas/chemical carriers; fully refrigerated carriers; LNG carriers; and CNG carriers. Gas carriers transport a wide selection of cargoes, including butadiene, ethylene, LPG, LNG, CNG, propylene, and chemical gases such as ammonia, vinyl chloride, ethylene oxide, propylene oxide, and chlorine. Most gas carriers are built in South Korea (Daewoo Shipbuilding & Marine, Hyundai Heavy Industries, Hyundai Mipo, Hyundai Samho Heavy Industries), China (Jiangnan), and Japan (Kawasaki Shipbuilding Corporation, and Mitsubishi Heavy Industries). Damen Shipyard in the Netherlands has built a small number of new buildings. In this chapter, we will briefly examine each of these types of vessels, as well as the gas carrier codes, cargo containment systems, and the hazards and health effects associated with gas carriers (Figure 11.1).

Figure 11.1 Typical type gas carrier *Marshal Vasil Evskiy*.

DOI: 10.1201/9781003342366-14

TYPES OF GAS CARRIERS

There are six types of gas carriers in operation: fully pressurised gas carriers, semi-pressurised gas carriers, ethylene and gas/chemical carriers, fully refrigerated ships, LNG carriers, and CNG carriers.

(1) *Fully pressurised gas carriers*. The seaborne transport of liquefied gases began in 1934 when a major international company put two combined oil/LPG tankers into operation. The ships, which were oil tankers, had been converted by fitting small, riveted, pressure vessels for the carriage of LPG into cargo tank spaces. This enabled the transport over long distances of substantial volumes of an oil refinery byproduct that had distinct advantages as a domestic and commercial fuel. LPG is not only odourless and non-toxic, but it also has a high calorific value and a low sulphur content, making it exceptionally clean and efficient when burnt. Today, most fully pressurised ocean-going LPG carriers are fitted with two or three horizontal, cylindrical, or spherical cargo tanks and have typical capacities ranging between 20,000 and 90,000 cu/m and length overall ranging from 140 to 229 m (459 ft–751 ft). New LPG carriers are designed with a dual-fuel propulsion system possessing the ability to use LPG or diesel fuel on a selective basis. Fully pressurised ships are still constructed in numbers and represent a cost-effective and straightforward way of moving LPG to and from smaller gas terminals.

(2) *Semi-pressurised ships*. Semi-pressurised ships carry gases in a semi-pressurised or a semi-refrigerated state. This approach provides flexibility, as these carriers can load or discharge at both refrigerated and pressurised storage facilities. Semi-pressurised and semi-refrigerated carriers incorporate cylindrical, spherical, or bi-lobe shaped tanks carrying propane under pressure of 8.5 kg/cm^2 (or 121 psi), at a temperature of –10°C (14°F).

(3) *Ethylene and gas/Chemical carriers*. Ethylene carriers are the most sophisticated type of gas carrier as they can carry not only most other liquefied gas cargoes but also ethylene at its atmospheric boiling point of –104°C (–155°F). These ships feature cylindrical, insulated, stainless steel cargo tanks able to accommodate cargoes up to a maximum specific gravity of 1.8 at temperatures ranging from a minimum of –104°C to a maximum of +80°C (176°F) and at a maximum tank pressure of 4 bar.

(4) *Fully refrigerated ships*. Fully refrigerated ships are built to carry liquefied gases at a low temperature and atmospheric pressure between terminals equipped with fully refrigerated storage tanks. However, discharge through a booster pump and cargo heater makes it possible to discharge to pressurised tanks as well. The first purpose-built, LPG tanker was the *MT Rasmus Tholstrup*, which was built in Sweden

in accordance with a Danish design. Prismatic tanks enabled the ship's cargo-carrying capacity to be maximised, thus making fully refrigerated ships particularly well suited for carrying large volumes of cargo such as LPG, ammonia, and vinyl chloride over substantial distances. Today, fully refrigerated ships range in capacity from 20,000 to 100,000 m^3 (710,000–3,530,000 cu/ft). LPG carriers in the 50,000–80,000 m^3 (1,800,000–2,800,000 cu/ft) size range are often referred to as VLGCs (very large gas carriers). Although LNG carriers are often larger in terms of cubic capacity, this term is normally only applied to fully refrigerated LPG carriers. The main type of cargo containment system used on board modern fully refrigerated ships is independent tanks with rigid foam insulation. The insulation used is most commonly polyurethane foam. Older ships may have independent tanks with loosely filled perlite insulation. In the past, there have been a few fully refrigerated ships built with semi-membrane or integral tanks and internal insulation tanks, but these systems attracted only minimal interest from ship owners.

(5) *Liquefied natural gas (LNG) carriers*. Many LNG carriers have between 125,000 and 135,000 m^3 (4,400,000 and 4,800,000 cu/ft) of carrying capacity, although from 1994, LNG carriers have been progressively smaller with an average of 18,000–19,000 m^3 (640,000 and 670,000 cu/ft).

(6) *CNG carriers*. CNG carriers are designed for transportation of natural gas under high pressure. CNG carrier technology relies on high pressure, typically over 250 bar (2900 psi), to increase the density of the gas and maximise the possible commercial payload. CNG carriers are economical for medium-distance marine transport and rely on the adoption of suitable pressure vessels to store CNG during transport and on the use of suitable loading and unloading compressors to receive the CNG at the loading terminal and to deliver the CNG at the unloading terminal.

GAS CARRIER CODES

The gas carrier codes, developed by the IMO in London, apply to all gas carriers regardless of their size. There are three gas carrier codes, the IGC Code, the *GC Code*, and the existing ship code, which are outlined in the following sections.

Gas carriers built after June 1986 (the IGC code)

The *IGC Code*, which applies to new gas carriers (built after 30 June 1986), is the *International Code for the Construction and Equipment of Ships Carrying Liquefied Gases in Bulk*. In summary, this Code is known as the *IGC Code*. The *IGC Code*, under amendments to The *International Convention for the Safety of Life at Sea* (SOLAS), is mandatory for all new ships. As a proof that a ship complies with the Code, an International Certificate of Fitness for the Carriage of Liquefied Gases in Bulk must be transported on board. In 1993, the *IGC Code* was amended, with new rules coming into effect on 1 July 1994. Ships, on which construction started on or after 1 October 1994, must apply the amended version of the Code, but ships built earlier than 1 July 1994 may continue to comply with the previous edition of the *IGC Code*.

Gas carriers built between 1976 and 1986 (the GC code)

The regulations covering gas carriers built after 1976 but before July 1986 are included in the *Code for the Construction and Equipment of Ships Carrying Liquefied Gases in Bulk*. It is known as the *Gas Carrier Code* or GC Code in short. Since 1975, the IMO has approved four sets of amendments to the *GC Code*. The latest amendment was adopted in June 1993. The amendments have not necessarily been agreed upon by every signing government. Although the Code is not mandatory, many countries have implemented it into their national laws. Accordingly, most charterers expect such ships to meet the Code standards and, as proof of this, to carry on board a Certificate of Fitness for the Carriage of Liquefied Gases in Bulk.

Gas carriers built before 1977 (the existing ship code)

The regulations covering gas carriers built before 1977 are contained in the *Code for Existing Ships Carrying Liquefied Gases in Bulk*. The content is like the *GC Code*, though less extensive. The *Existing Ship Code* was completed in 1976 after the GC Code had been written. It therefore summarises current shipbuilding practice at that time. It remains an IMO recommendation for all gas carriers in this older fleet of ships. The Code is not mandatory but is applied by some countries for ship registration and in other countries as a necessary fulfilment prior to port entry. Accordingly, many ships of this age are required by charterers to meet Code standards and to transport on board a Certificate of Fitness for the Carriage of Liquefied Gases in Bulk.

HAZARDS AND HEALTH EFFECTS

Toxicity

Vinyl chloride, a product commonly transported on gas carriers, is a known human carcinogen. It is especially linked to liver cancer. It is not only dangerous when inhaled but can also be absorbed by the skin. Skin irritation and watering of the eyes indicate that dangerous levels of vinyl chloride may be present in the atmosphere. Caution must be exercised while dealing with such cargoes, including the use of precautions such as chemical suits, self-contained breathing apparatus (SCBA), and gas-tight goggles. Similar safety precautions should be taken when carrying other toxic cargoes including chlorine and ammonia.

Hazards of ammonia

Ammonia is an extremely hazardous chemical. Exposure to more than 2,000 parts per million (ppm) can be fatal within 30 minutes, whereas exposure to 6,000 ppm is fatal within minutes. Exposure to as much as 10,000 ppm is fatal instantly and intolerable to unprotected skin. Anhydrous ammonia is not dangerous when managed properly, but when not managed carefully, it can be extremely dangerous. It is not as combustible as many other products that are used and managed every day. However, high concentrations of ammonia gas can burn and requires precautions to avoid the outbreak of fire. Mild ammonia exposure can cause irritation to the eyes, nose, and lung tissues. Prolonged breathing of ammonia vapours can cause suffocation. When substantial amounts are inhaled, the throat swells shut, causing the casualty to suffocate. Exposure to vapours or liquid can cause temporary and permanent blindness. It is the water-absorbing nature of anhydrous ammonia that causes the greatest injury (especially to the eyes, nose, throat, and lungs), which can cause permanent damage. It is a colourless gas at atmospheric pressure and normal temperature, but under pressure readily changes into a liquid. Anhydrous ammonia has a high affinity for water.

Anhydrous ammonia is a hygroscopic compound, which means it will seek a moisture source. This includes unprotected skin, which is composed of approximately 90% water. When a human body is exposed to anhydrous ammonia, the chemical freeze burns its way into the skin, eyes, or lungs. This attraction places the eyes, lungs, and skin at the greatest risk because of their high moisture content. Caustic burns can result when the anhydrous ammonia dissolves into body tissue. Most deaths from anhydrous ammonia are caused by severe damage to the throat and lungs from a direct blast to the face. An additional concern is the low boiling point of anhydrous ammonia. The chemical freezes on contact at room temperature. This causes burns like, but more severe than, those caused by dry ice. If exposed to severe cold, the flesh will become frozen. At first, the skin will become red (but turn subsequently white); the affected area is painless, but hard to touch, and if left untreated, the flesh will die and may become gangrenous. The human eye is a complex organ made up of about 80% water. Ammonia under pressure can cause extensive, almost immediate eye damage on contact. The ammonia extracts the fluid and destroys the eye cells and tissue in minutes. Draining ammonia into the sea whilst pre-cooling the hard-arm or during disconnection operations is not environment-friendly and should be avoided.

Even a small quantity of ammonia [as low as 0.45 mg/l (1.6×10–8 lb/cu in) (LC50)] is hazardous to aquatic life, including edible species of fish such as cod and salmon. The entry of contaminated seafood into the human food chain presents considerable problems.

Flammability

All cargo vapours are flammable. When ignition occurs, it is not the liquid which burns but the evolved vapour. Flameless explosions, which result from cold cargo liquid coming into sudden contact with water, do not release much energy. Pool fires, which are the result of a leaked pool of cargo liquid catching fire, and jet fires, which are the result of the leak catching fire, are serious hazards. Flash fires occur when there is a leak and do not ignite immediately but after the vapours travel some distance downwind, at which point they ignite. This is particularly dangerous on ships. Vapour cloud explosions and boiling liquid expanding vapour explosions (BLEVE) are the most dangerous hazards on gas carriers.

Frostbite

On gas carriers, cargoes are carried at extremely low temperatures ranging from 0°C to −163°C (32°F to −261°F). This means that frostbite caused by exposure of skin to the cold vapours or liquid is a very real hazard.

Asphyxia

Asphyxia occurs when the blood cannot bring a sufficient supply of oxygen to the brain. A person affected by asphyxia may experience headaches, dizziness, and an inability to concentrate, followed by loss of consciousness. In sufficient concentrations, any vapour may cause asphyxiation, whether toxic or not.

Spillage

Compared with oil, there is less industry concern over the spillage of cargoes from LNG carrying vessels as the gas will quickly vapourise. The LNG sector is known for having a good safety record regarding cargo loss. By 2004, there had been close to 80,000 loaded port transits of LNG carriers with no loss of containment failure. An analysis of several spherical carriers showed that the vessels can withstand a 90-degree side-on collision with another similar LNG carrier at 6.6 knots (50% of normal port speed) with no loss of LNG cargo integrity. This drops to 1.7 knots for a fully loaded 300,000 dwt oil tanker collision into an LNG carrier. The report also notes that such collisions are rare. In 2004, HAZID performed a risk assessment of an LNG spill. Taking account of precautions, training, regulations, and technology changes over time, HAZID calculates that the likelihood of an LNG spill is approximately 1 in 100,000 trips. If the tank integrity of a LNG transport is compromised, there is a risk that the natural gas contained within could ignite, causing either an explosion or fire.

In the first of two chapters examining the role and function of gas carrier type vessels, we have looked at the type of ship, which transports gas in a vapour state. In the second part of the two chapters, we will look at how gas is transported in liquid form as LNG.

Chapter 12

LNG carriers

As discussed previously, the LNG carrier is a type of ship designed specifically for transporting LNG. The first LNG carrier, the *Methane Pioneer* (5,034 dwt), left the Calcasieu River on the Louisiana Gulf coast on 25 January 1959. Carrying the world's first ocean cargo of LNG, it sailed to the United Kingdom where the cargo was discharged. Subsequent expansion of that trade brought about a large expansion of the LNG fleet where today giant LNG ships carrying up to 266,000 m^3 (9,400,000 cu ft) are operating worldwide. The success of the specially modified C1-M-AV1-type standard ship *Normarti*, which was renamed *Methane Pioneer*, caused the Gas Council and Conch International Methane Ltd. to order two purpose-built LNG carriers to be constructed: the *Methane Princess* and *Methane Progress*. Both ships were fitted with Conch-independent aluminium cargo tanks, entering the Algerian LNG trade in 1964. These ships had a capacity of 27,000 m^3 (950,000 cu/ft). In the late 1960s, opportunities arose to export LNG from Alaska to Japan. In 1969, trade between TEPCO and Tokyo Gas was initiated. Two ships, *Polar Alaska* and *Arctic Tokyo*, each with a capacity of 71,500 m^3 (2,520,000 cu/ft), were built in Sweden. In the early 1970s, the US government encouraged US shipyards to build LNG carriers, with a total of 16 LNG ships built in US shipyards. By the late 1970s and early 1980s, the prospect of Arctic-capable LNG ships brought about a renewed interest in the development of LNG carrier technology and designs. With the increase in cargo capacity rising to approximately 143,000m^3 (5,000,000 cu/ft), new tank designs were developed starting with the *Moss Rosenberg*, to the *Technigaz Mark III*, and Gaztransport's *No.96*. Over recent years, the size and capacity of LNG carriers have increased exponentially. In 2005, Qatargas pioneered the development of two new classes of LNG carriers: the Q-Flex and Q-Max. Each ship has a cargo capacity of between 210,000 and 266,000 m^3 (7,400,000 and 9,400,000 cu ft) and is equipped with re-liquefaction plants. At the same time, the development of small-scale LNG bunker carriers has also increased, in line with the maritime industry's move towards greener energy. Some models, such as the Damen LGC 3,000, are compact enough to fit beneath the life rafts of cruise ships and ROPAX vessels (Figure 12.1).

DOI: 10.1201/9781003342366-15

Figure 12.1 Typical LNG carrier *Gas Spirit*.

In November 2018, South Korean shipbuilders locked in three years' worth of large-scale LNG carrier contracts totalling more than 50 orders with a list value of over US$9 billion. This is equal to 78% of LNG-related shipbuilding contracts, with 14% going to Japanese shipbuilders and 8% going to Chinese shipyards. These new buildings would boost the global LNG fleet by some 10%. Of the global fleet, historically, about two-thirds of LNG carriers have been built in South Korea, 22% by Japan, 7% by China, and the remainder by a combination of France, Spain, and the United States. South Korea's success stems from innovation and price points. For example, South Korean shipbuilders introduced the first ice-breaker type LNG vessels. In 2017, Daewoo Shipbuilding & Marine Engineering delivered the *Christophe de Margerie*, the world's first icebreaking LNG tanker, with a dwt of 80,200 metric tonnes, with the option to build a further 14 new buildings on contract. Her capacity of 172,600 m^3 (6,100,000 cu/ft) is equal to the entire gas consumption of Sweden for a month. She completed her first revenue voyage from Norway via the Northern Sea Route in the Arctic Ocean to South Korea. According to SIGTTO data, in 2019, there were 154 LNG carriers on order, and 584 operating LNG carriers worldwide.

A typical LNG carrier has four to six tanks located along the centreline of the vessel. Surrounding the tanks is a combination of ballast tanks, cofferdams, and void spaces; this effectively gives the vessel a double-hull type design. Inside each tank, there are typically three submerged pumps. There are two main cargo pumps, which are used in cargo discharge operations and a much smaller pump, which is referred to as the spray pump. The spray pump is used for either pumping out liquid LNG to be used as fuel (via a vapouriser), or for cooling down cargo tanks. It can also be used for 'stripping' out the last of the cargo in discharge operations. All these pumps are contained within what is known as the pump tower, which hangs from the top of the tank and runs the entire depth of the tank. The pump tower also contains the tank gauging system and the tank filling line, all of which are located near the bottom of the tank. In membrane-type vessels, there is also an empty pipe with a spring-loaded foot valve that can be opened by weight or pressure. This is the emergency pump tower. In the event, both main cargo pumps fail the top can be removed from this pipe and an emergency cargo pump lowered down to the bottom of the pipe. The top is replaced on the column and then the pump is allowed to push down on the foot valve and open it. The cargo can then be pumped out. All cargo pumps discharge into a common pipe, which runs along the deck of the vessel; it branches off to either side of the vessel to the cargo manifolds, which are used for loading or discharging. All cargo tank vapour spaces are linked via a vapour header, which runs parallel to the cargo header. This also has connections to the sides of the ship next to the loading and discharging manifolds.

CARGO HANDLING AND CYCLE

A typical cargo cycle starts with the tanks in a 'gas-free' condition, meaning the tanks are full of air, which allows maintenance on the tank and pumps. Cargo cannot be loaded directly into the tank, as the presence of oxygen would create an explosive atmospheric condition within the tank, and the rapid temperature change caused by loading LNG at –162°C (–260°F) could damage the tanks. First, the tank must be 'inerted' to eliminate the risk of explosion. An inert gas plant burns diesel in air to produce a mixture of gases (typically less than 5% O^2 and about 13% CO^2 and N^2). This is blown into the tanks until the oxygen level is below 4%. Next, the vessel goes into port to 'gas-up' and 'cool-down', as it is still unsafe to load directly into the tank. The CO^2 will freeze and damage the pumps and the cold shock could damage the tank's pump column. LNG is brought onto the vessel and taken along the spray line to the main vapouriser, which boils off the liquid into a gas. This is then warmed up to roughly 20°C (68°F) in the gas heaters and then blown into the tanks to displace the 'inert gas'. This continues until all the CO^2 is removed from the tanks. Initially, the inert gas is vented to the atmosphere. Once the hydrocarbon content reaches 5% (the lower flammability range of methane), the inert gas is redirected to shore via a pipeline and manifold connection by high-duty compressors. The shore terminal then burns this vapour to avoid the danger of having copious amounts of hydrocarbons present which may explode. Now, the vessel can be safely 'gassed up' and warmed. Currently, the tanks are still at ambient temperature and are full of methane. The next stage in the process is the 'cool-down'. LNG is sprayed into the tanks via spray heads, which vapourises and starts to cool the tank. The excess gas is again blown ashore to be re-liquified or burned at a flare stack. Once the tanks reach about –140°C (–220°F), the tanks are ready to bulk load. Bulk loading starts and liquid LNG is pumped from the storage tanks ashore into the vessel tanks. Displaced gas is blown ashore by the high-duty compressors. Loading continues until the tanks are typically 98.5% full. This allows for thermal expansion and contraction of the cargo.

The vessel can now proceed to the discharge port. During passage, various boil-off management strategies may be used to manage the cargo. Boil-off gas can be burned in the boilers to provide propulsion, or it can be re-liquefied and returned to the cargo tanks, depending on the design of the vessel. Once the vessel reaches the discharge port, the cargo is pumped ashore using the cargo pumps. As the tank empties, the vapour space is filled with either gas from ashore or by vapourising some of the cargo in the cargo vapouriser. The vessel will be either pumped out as much as possible, with the last residues being pumped out with spray pumps, or else some cargo may be retained on board as a 'heel'. It is a customary practice to keep between 5% and 10% of the cargo after discharge in one tank. This is referred to as the heel and is used to cool down the remaining tanks that have no heel before loading. This must be done gradually, otherwise the tanks will suffer cold shock if loaded directly into warm tanks. Cool-down can take 20 hours on *Moss class* vessels and 10–12 hours on membrane-type vessels, so carrying a heel allows cool-down to be done before the vessel reaches port. This provides a considerable time (and therefore cost) saving. If all the cargo is pumped ashore, then on the ballast passage, the tanks will warm up to ambient temperature, returning the vessel to a gassed up and warm state. This means the vessel must be cooled again for loading. If the vessel is to return to a gas-free state, the tanks must be warmed up by using the gas heaters to circulate warm gas. Once the tanks are warmed up, the inert gas plant is used to remove the methane from the tanks. Once the tanks are methane free, the inert gas plant is switched to dry air production, which is used to remove all the inert gas from the tanks until they have a safe working atmosphere.

CARGO CONTAINMENT

On most modern LNG carriers, there are four containment systems in use. Two of the designs are of the self-supporting type, while the other two are of the membrane type. Today, the French naval engineering company Gaztransport & Technigaz (GTT) owns the patents. There is a trend towards the use of the two different membrane types instead of the self-supporting storage systems. This is most likely because prismatic membrane tanks use the hull shape more efficiently and thus have less void space between the cargo and ballast tanks. As a result of this, the *Moss class* design, when compared with the membrane design of equal capacity, is more expensive to transit the Suez Canal. However, self-supporting tanks are more robust and have greater resistance to sloshing forces. This means that they will be considered in the future for offshore storage where inclement weather conditions will be a significant factor.

Moss tanks (spherical IMO type B LNG tanks)

Named after the Norwegian company, Moss Maritime, which designed them, the IMO type B LNG tanks are spherical in shape. Most *Moss class* vessels have four or five tanks. The outside of the tanks has a thick layer of foam insulation that is either fitted in panels or in contemporary designs wound round the tank. Placed over this insulation is a thin layer of 'tinfoil', which allows the insulation to be kept dry with a nitrogen atmosphere. This atmosphere is constantly checked for any methane that would indicate a leak of the tank. Furthermore, the outside of the tank is checked at three-monthly intervals for any cold spots that would indicate a breakdown in the insulation. The tank is supported around its circumference by the equatorial ring, which is supported by a large circular skirt, known as a data-couple. This is constructed from a unique combination of aluminium and steel, which takes the weight of the tank down to the ship's structure. This skirt allows the tank to expand and contract during cool-down and warm-up operations. During cool-down or warm-up, the tank can expand or contract by about 60 cm (24 in). Because of this expansion and contraction, the piping into the tank comes in from the top and is connected to the ship's lines via flexible bellows. Inside each tank are a set of spray heads. These heads are mounted around the equatorial ring and are used to spray LNG onto the tank walls to reduce the temperature. Tanks normally have a working pressure of up to 22 kPa (3.2 psi) (0.22 bar), but this can be increased for an emergency discharge. If both main pumps fail to remove the cargo, the tank's safety valves are adjusted to lift at 100 kPa (1 bar). The filling line, which goes to the bottom of the tank, is opened together with the filling lines of the other tanks on board. The pressure is then raised in the tank with the defective pumps. This pushes the cargo into the other tanks where it can be safely pumped out (Figure 12.2).

IHI (prismatic IMO type B LNG tanks)

Designed by Ishikawajima-Harima Heavy Industries, the self-supporting prismatic type B (SPB) tank is currently employed on only two vessels. Type B tanks limit sloshing problems, an improvement over the membrane LNG carrier tanks, which may break due to sloshing impact, therein impacting the ship's hull. This is also of prime relevance for FPSO LNG (or FLNG) vessels.

Figure 12.2 MOSS gas carrier *Arctic Princess*, Hammerfest, Norway.

TGZ Mark III

Designed by Technigaz, these tanks are of the membrane type. The membrane consists of stainless steel with 'waffles' to absorb thermal contractions when the tank is cooled down. The primary barrier, made of corrugated stainless steel [with a thickness of about 1.2 mm (0.047 in)], is in direct contact with the cargo liquid (or vapour in empty tank condition). This is followed by a primary insulation, which in turn is covered by a secondary barrier made of a material called 'triplex'. Triplex is a type of metal foil, which is sandwiched between glass wool sheets and firmly compressed together. This is covered by a secondary layer of insulation, which in turn is supported by the ship's hull structure from the outside. From the inside of the tank outwards, the layers are:

- LNG cargo;
- Primary barrier of 1.2 mm thick corrugated/waffled 304L stainless steel;
- Primary insulation (also called the inter barrier space);
- Secondary barrier within a triplex membrane;
- Secondary insulation (also called the insulation space);
- Ship's hull structure.

GT96

Designed by Gaztransport, the tanks consist of a primary and secondary thin membrane made of Invar, which has no thermal contraction. The insulation is made from plywood boxes filled with perlite. These are continuously flushed with nitrogen gas. The integrity of both membranes is permanently monitored for the detection of hydrocarbon vapours within the nitrogen.

CS1

CS1 stands for Combined System Number One. It was designed by the now merged Gaztransport and Technigaz companies and consists of the best components of both the Mk III and GT96 systems. The primary barrier is made of Invar [0.7 mm (0.028 in)], and a secondary layer consisting of Triplex. The primary and secondary insulation consists of polyurethane foam panels. Although three vessels with CS1 technology were built, the established shipyards have decided to maintain production of the separate Mk III and GT96 types.

RE-LIQUEFACTION AND BOIL OFF

To facilitate transport, the natural gas must be cooled down to approximately −163°C (−261°F) at atmospheric pressure. At this temperature, the gas condenses into a liquid. The tanks on board an LNG carrier effectively function as giant thermoses to keep the liquid gas cold during storage. As no insulation is perfect, the liquid must be constantly boiled during the voyage. According to WGI, on a typical voyage, an estimated 0.1–0.25% of the cargo converts to gas each day, depending on the efficiency of the insulation and the roughness of the passage. In a typical 20-day passage, anywhere from 2% to 6% of the total volume of LNG originally loaded may be lost. Most LNG carriers are powered by steam turbines connected to marine boilers. These boilers are dual fuel and can run on either methane, oil, or a combination of both. The gas produced through boil off is traditionally diverted to the boilers and used as a fuel for the vessel. Before this gas is used in the boilers, it must be warmed up to 20°C by using the gas heaters. The gas is either fed into the boiler by tank pressure or it is increased in pressure by the low duty compressors. What fuel the vessel runs on is dependent on many factors, which include the length of the passage, any requirement to carry heel for cooldown, the prevailing price of oil versus the price of LNG, and port requirements for cleaner exhaust. There are three basic modes available:

- Minimum boil-off/maximum oil. In this mode, tank pressures are kept high to reduce boil off to a minimum and most of the energy comes from fuel oil. This maximises the amount of LNG delivered but causes tank temperatures to rise due to lack of evaporation. The high cargo temperatures can cause storage and offloading problems.

- Maximum boil-off/minimum oil. With this mode the tank pressures are kept low. This means there is greater boil-off, but a large amount of fuel oil is still used. This decreases the amount of LNG delivered but the cargo will be delivered cold, which many ports prefer.
- 100% gas. Tank pressures are kept at a similar level to maximum boil off but as this may not be enough to supply the boilers needs, additional LNG must be 'forced' to vapourise. A small pump in one tank supplies LNG to the forcing vapouriser, where it is warmed and vapourised back into a gas that is usable in the boilers. In this mode, no fuel oil is used.

Advances in reliquefication plant technology has allowed boil-off to be reliquefied and returned to the tanks. Because of this, vessel designers have been able to introduce more efficient slow-speed diesel engines (previously most LNG carriers were steam turbine-powered). Notable exceptions are the LNG carrier *Havfru* (built as *Venator* in 1973), which originally had dual fuel diesel engines, and her sister-ship *Century* (built as *Lucian* in 1974), also built with dual fuel gas turbines before being converted to a diesel engine system in 1982. Today, vessels using Dual, or Tri-Fuel Diesel Electric (DFDE/TFDE, respectively) propulsion systems are in service. Over the past couple of years, there has been increasing interest in a return to propulsion by boil-off gas. This is a result of the IMO 2020 anti-pollution regulation that bans the use of marine fuel oil with a sulphur content greater than 0.5% on ships not fitted with a flue-gas scrubbing plant. Space constraints and safety issues typically prevent the installation of such equipment on LNG carriers, forcing them to abandon the use of the lower-cost, high-sulphur fuel oil and switch to low-sulphur fuels that cost more and are in shorter supply. In these circumstances, boil-off gas may become a more attractive option.

In the next chapter, we will examine the final vessel type in Part II, the oil tanker.

Chapter 13

Oil tankers and product carriers

Ships that carry liquid cargoes (either as a crude or refined product) are called oil tankers or petroleum tankers. There are two basic types of oil tankers: crude tankers and product tankers. Crude tankers move enormous quantities of unrefined crude oil from its point of extraction to oil refineries. Product tankers, which are much smaller, are designed to move refined products from refineries to offloading points near consuming markets. Oil tankers are often classified by their size as well as their occupation. The size classes range from inland or coastal tankers of a few thousand metric tonnes dwt to the mammoth ultra large crude carriers (ULCCs) of 550,000 dwt. Tankers move approximately two billion metric tonnes (2.2 billion short tonnes) of oil every year. Second only to pipelines in terms of efficiency, the average cost of transporting crude oil by tanker amounts to only US$5 to $8 per cubic metre (US$0.02 to US$0.03 per US gallon). Some specialised types of oil tankers have also evolved. One of these is the naval replenishment oiler, a tanker which can fuel a moving vessel. Combination ore-bulk-oil carriers and permanently moored floating storage units are two other variations of the standard oil tanker design. Oil tankers have been involved in several environmentally damaging and high-profile oil spills. As a result, they are subject to stringent design and operational regulations (Figure 13.1).

Figure 13.1 Typical oil tanker *MT Seavigour.*

DOI: 10.1201/9781003342366-16

DEVELOPMENT OF THE OIL TANKER

The technology of oil transportation has evolved alongside the oil industry. Although human use of oil reaches as far back as prehistory, the first modern commercial exploitation dates to James Young's manufacture of paraffin in 1850. In the early 1850s, oil began to be exported from Upper Burma, which was then a British colony. The oil was moved in earthenware vessels to the riverbank where it was poured into boat holds for transportation to Britain. In the 1860s, Pennsylvanian oil fields became a major supplier of oil, and a centre of innovation after Edwin Drake struck oil near Titusville, Pennsylvania. Break-bulk boats and barges were used to transport this oil in 40 US gallon (150 l) wooden barrels. Transport by barrel had several issues. First, was the weight. An empty barrel weighs about 29 kg (64 lbs) or about 20% of the weight of the full barrel. Second, was the barrel's tendency to leak. This meant not only was valuable product being lost through ullage, but leaks also presented significant slip and fire hazards on board. Third, and most significant, was the cost of purchasing the barrels in the first place. In the early days of the Russian oil industry, barrels accounted for almost half of all petroleum production costs, as it was not unusual for each barrel to be used only once. In 1863, two sail-driven tankers were built on the River Tyne, England. By 1871, the Pennsylvania oil fields were making limited use of oil tank barges and cylindrical railroad tank cars like those in use today. The two English ships were followed in 1873 by the first oil-tank steamer, *Vaderland* (Fatherland), which was built by the Palmers Shipbuilding and Iron Company for Belgian owners. The vessel's use was curtailed by American and Belgian authorities citing safety concerns.

The modern oil tanker was developed between 1877 and 1885. In 1876, the brothers of the Swedish pioneer Alfred Nobel, Ludvig and Robert Nobel, founded the Branobel Company (short for Brothers Nobel) in Baku, Azerbaijan. During the late 19th century, Branobel would become one of the largest oil companies in the world. Ludvig Nobel was a pioneer in the development of early-age oil tankers. He first experimented with carrying oil in bulk on single-hulled barges. When he realised this was not terribly efficient, Nobel turned his attention to self-propelled tankships, although here too he encountered several problems. The major concern was how to keep the cargo and fumes away from the engine room to avoid fires. Other challenges included allowing for the cargo to expand and contract caused by variations in atmospheric temperature and developing means of ventilating the cargo tanks. Based on a series of technological advancements and enhancements, the first successful oil tanker to be built was the *Zoroaster*, which carried its 246 metric tonnes (242 long tonnes) of kerosene cargo in two iron tanks joined together by pipes. One tank was positioned forward of the midships engine room, and the other tank was positioned aft. The ship also featured a series of 21 vertical watertight compartments, which provided extra buoyancy. The ship had a length overall of 56 m (184 ft), a beam of 8.2 m (27 ft), and a draught of 2.7 m (9 ft). Unlike later Nobel tankers, the *Zoroaster* design was built small enough to sail from Sweden to the Caspian via the Baltic Sea, Lake Ladoga, Lake Onega, the Rybinsk and Mariinsk Canals, and then up the Volga River. In 1883, oil tanker design took a large step forward. Whilst working for the Nobel Company, British engineer Colonel Henry F. Swan designed a revolutionary tank design. Instead of having one or two large holds, Swan's design consisted of several holds spanning the entire beam of the ship. These holds were further subdivided into port and starboard sections by a longitudinal bulkhead. Whereas earlier tanker designs suffered from stability problems caused by free surface effect (where oil sloshing from side to side could cause a ship to capsize), the inclusion of longitudinal bulkheads, which divided the ship's storage space into smaller tanks, virtually eliminated the problems caused by free-surface effect. This approach, almost universal today, was first used by Swan in the Nobel tankers *Blesk*, *Lumen*, and *Lux*. Some naval historians point to the *Glückauf*, another design of Colonel Swan that was built in England, as being the first modern oil tanker. This ship adopted the best practices from earlier tanker designs to create the prototype for all subsequent vessels of this type. It was the first dedicated steam-driven ocean-going tanker in the world and was the first ship in which oil could be pumped directly into the vessel hull instead of being loaded in barrels or drums. It was also the first tanker to be constructed with a horizontal bulkhead. The *Glückauf's* other features included cargo valves operable from the deck, cargo main piping, a vapour line, cofferdams for added safety, and the ability to fill a ballast tank with seawater when empty of cargo. The ship was built in England before being purchased by Wilhelm Anton Riedemann, an agent for the Standard Oil Company along with several of her sister ships. The *Glückauf* was lost in 1893 after grounding in heavy fog.

The 1880s also saw the beginnings of the Asian oil trade. The concept of shipping Russian oil to the Far East via the Suez Canal was the brainchild of two men: importer Marcus Samuel (founder of the Shell Transport and Trading Company) and shipowner/broker Fred Lane. Previous attempts to ship oil through the Suez Canal were refused by the Suez Canal Company as being too dangerous. Not accepting the impasse, Samuel and Lane tried a different tactic and approached the Suez Canal Company directly, asking for the specifications of a tanker, which would be permitted entry through the canal. Armed with the canal company's specific specifications, Samuel ordered three tankers from the northern English shipbuilder William Gray & Company. These ships, the *Murex*, *Conch*, and *Clam*, each had a capacity of 5,010 long tonnes dwt, and formed the Tank Syndicate, forerunner of today's Royal Dutch Shell Company. With facilities in Jakarta (Indonesia), Singapore, Bangkok (Thailand), Saigon (Vietnam), Hong Kong, Shanghai

(China), and Kobe (Japan), the fledgling Shell company was ready to become Standard Oil's first major challenger in the Asian market. On 24 August 1892, Murex became the first tanker to pass through the Suez Canal. By the time the Shell Trading and Transport Company merged with Royal Dutch Petroleum in 1907, the company had a fleet of 34 steam-driven oil tankers compared with Standard Oil's four case-oil steamers and 16 sailing tankers.

Until 1956, tankers were designed to be able to navigate the Suez Canal. This size restriction became much less of a priority after the closing of the canal during the Suez Crisis of 1956. Forced to move oil around the Cape of Good Hope, shipowners quickly realised that bigger tankers were the key to more efficient transport. While a tanker of an era of typical World War II measured 162 m (532 ft) long and had a capacity of 16,500 dwt, the ultralarge crude carriers (ULCC) built around the 1970s were over 400 m (1,300 ft) long and had a capacity of 500,000 dwt. Several factors encouraged this growth. Increasing hostilities in the Middle East, which interrupted traffic through the Suez Canal contributed, as did the mass nationalisation of Middle East oil refineries. Fierce competition among shipowners also played contributing part. Aside from these factors, is a simple economic advantage: the larger an oil tanker is, the more cheaply it can move crude oil, and the better it can help meet growing demands for oil. In 1955, the world saw the launch of the then-largest supertanker, the *SS Spyros Niarchos*, which was built by the Vickers Armstrong shipyard in England for the Greek shipping magnate Stavros Niachos. The *SS Spyros Niarchos* had a gross tonnage of 30,708 and 47,500 long tonnes dwt. The *SS Spyros Niarchos* would retain her title as the largest vessel afloat until only 1958 when the American shipowner Daniel K. Ludwig broke the record for 100,000 long tonnes of heavy displacement. The *Universe Apollo* weighed in at a cool 104,500 long tonnes, itself a 23% increase from the previous record-holder, the *Universe Leader*, which also sailed under the Ludwig House Flag. The decisive step in the evolution of the supertanker (in terms of size) came in 1979 when the world's largest supertanker was at the Oppama shipyard in Japan by Sumitomo Heavy Industries, Ltd., and aptly named the *Seawise Giant*. This ship was built with a capacity of 564,763 dwt, a length overall of 458.45 m (1,504.1 ft), and a draught of 24.61 m (80.74 ft). In total, she had 46 tanks, 31,541 square metres (339,500 ft^3) of deck, and at full load draught, was prevented from navigating the English Channel. In 1989, the *Seawise Giant* was renamed the *Happy Giant*; renamed again in 1991 as the *Jahre Viking*; and then the *Knock Nevis* in 2004. In that year, she was converted into a permanently moored storage tanker and remained so until 2009 when she was sold for the final time; renamed *Mont* and finally scrapped in India in 2010. At the time *Knock Nevis* was scrapped, the world's largest oil tankers were the four *TI-class* sister ships, *TI Africa* (2002), *TI Europe* (2002), TI Asia

Figure 13.2 Aframax oil tanker *Ab Qaiq*.

(2003), and *TI Oceania* (2003). These vessels were the first ULCC supertankers to be built in almost 25 years. Boasting a hefty displacement of 67,591 light tonnes and 509,484 metric tonnes fully loaded, and a dwt tonnage of 441,893 and gross weight tonnage of 234,006, the TI class ships were the largest afloat but for the *Pioneering Spirit*, which is a crane ship. In 2009 and 2010, *TI Asia* and *TI Africa* were converted into FSOs, and in 2017, *TI Europe* was charted to the Norwegian oil company, Statoil, and converted into an FSO. In 2019, the last remaining ship of the *TI class*, the *TI Oceania*, was converted into an FSO and moored off the coast of Singapore (Figure 13.2).

Except for the pipeline, the tanker is the most cost-effective way to move oil today. Worldwide, tankers carry some two billion barrels (3.2×10^{11} litres) annually, and the cost of transportation by tanker amounts to only US$0.05 per gallon at the pump.

SIZE CATEGORIES

In 1954, the Shell Oil Company developed the 'average freight rate assessment' (AFRA) system. This system classified tankers by their assorted sizes. To make it an independent instrument, Shell consulted the London Tanker Brokers' Panel (LTBP). At first, they divided the groups as General Purpose for tankers under 25,000 metric tonnes dwt; Medium Range for ships between 25,000 and 45,000 dwt and Long Range for ships that were larger than 45,000 dwt. As tankers increased in size throughout the 1970s, the AFRA system underwent some revision. The system was developed for tax reasons, as the tax authorities wanted evidence that the internal billing records were correct. Before the New York Mercantile Exchange started trading crude oil futures in 1983, it was difficult to determine the exact price of oil, which could change with every contract. Shell and BP, the first companies to use the system, abandoned the AFRA system in 1983. They were later followed by the US oil companies. Despite this, the system is still used today as a flexible market scale, which takes typical routes and lots of 500,000 barrels (79,000 m^3). Oil tankers carry a wide range of hydrocarbon liquids ranging from crude oil to refined petroleum products. Their size is measured in dwt metric tonnes. Crude carriers are among the largest, ranging from 55,000 dwt Panamax-sized vessels to ultralarge crude carriers (ULCCs) of over 440,000 dwt smaller tankers, typically ranging from under 10,000 dwt to 80,000 dwt Panamax vessels, carry refined petroleum products, and are known as product tankers. The smallest tankers, with capacities under 10,000 dwt, work near coastal and inland waterways. Although they were considered such in the past, ships of the smaller Aframax and Suezmax classes are no longer regarded as super tankers. 'Supertankers' are the largest oil tankers and the largest man-made mobile structures. They include exceptionally large and ultralarge crude carriers (VLCCs and ULCCs, respectively) with capacities over 250,000 dwt. These ships can transport two million barrels (320,000m^3) of oil – the equivalent of 318,000 metric tonnes. By way of comparison, in 2009, the UK consumed about 1.6 million barrels (250,000m^3) of oil per day. The colossal ULCCs commissioned in the 1970s were the largest vessels ever built but have since been scrapped. A few newer ULCCs remain in service, none of which are more than 400 m (1,312 ft) long. Because of their size, super tankers often cannot enter port fully loaded. These ships can take on their cargo at offshore platforms and single-point moorings. At the other end of the journey, they often pump their cargo off to smaller tankers at designated lightering points in inshore waters. Super tanker routes are typically long, requiring them to stay at sea for extended periods, often around 70 days at a time (Table 13.1).

Table 13.1 Oil tanker size categories (2007 values)

Class	*Size in DWT*	*Class*	*Size in DWT*	*Newbuilding (US$ m)*	*Used (US$ m)*
General Purpose	10,000–24,999	Product Tanker	10,000–60,000	43	42.5
Medium Range	25,000–44,999	Panamax	60,000–80,000		
LR1	45,000–79,999	Aframax	80,000–120,000	60.7	58
LR 2	80,000–159,999	Suezmax	120,000–200,000		
VLCC	160,000–319,000	VLCC	200,000–320,000	120	116
ULCC	320,000–549,999	ULCC	320,000–550,000		

TANKER CHARTERING

The act of hiring a ship to carry cargo is called chartering. Tankers are hired by four types of charter agreements: the *voyage charter*, the *time charter, the bareboat charter*, and *contract of affreightment*. In a voyage charter, the charterer rents the vessel from the loading port to the discharge port. In a time charter, the vessel is hired for a set period, to perform voyages as the charterer directs. In a bareboat charter, the charterer acts as the ship's operator and manager, taking on responsibilities such as providing the crew and maintaining the vessel in a seaworthy condition. Finally, in a contract of affreightment or COA, the charterer specifies a total volume of cargo to be carried over a specific period and in specific sizes. For example, a COA could be specified as one million barrels (160,000 m^3) of JP-5 in a year's time in 25,000-barrel (4,000 m^3) shipments. A completed chartering contract is known as a charter party. One of the key aspects of any charter party is the freight rate, or the price specified for the carriage of cargo. The freight rate of a tanker charter party is specified in one of four ways: by a lump sum rate, by rate per tonne, by a time charter equivalent rate, or by the Worldscale rate. In a lump sum rate arrangement, a fixed price is negotiated for the delivery of a specified cargo, and the ship's owner/operator is responsible for paying all port costs and other voyage expenses. Rate per tonne arrangements is used mostly in chemical tanker chartering and differ from lump sum rates in that port costs and voyage expenses are paid by the charterer. Time charter arrangements specify a daily rate, and port costs and voyage expenses are also generally paid by the charterer. The Worldwide Tanker Normal Freight Scale, often referred to as Worldscale, is established and governed jointly by the Worldscale Associations of London and New York. Worldscale establishes a baseline price for carrying a metric tonne of product between any two ports worldwide. In Worldscale negotiations, operators and charterers determine a price based on a percentage of the Worldscale rate. The baseline rate is expressed as WS 100. If a given charter party settles on 85% of the Worldscale rate, it would be expressed as WS 85. Similarly, a charter party set at 125% of the Worldscale rate would be expressed as WS 125 (Table 13.2).

FLEET CHARACTERISTICS

The global total oil tanker dwt tonnage increased from 326.1 million dwt in 1970 to 960.0 million dwt in 2005. In 2005, oil tankers made up 36.9% of the world's fleet in terms of dwt tonnage. Combined, the dwt tonnage of oil tankers and bulk carriers represents 72.9% of the world's merchant shipping fleet. In same year, some 2.42 billion metric tonnes of oil were shipped by tanker. About 76.7% of this was crude oil, with the remainder consisting of refined petroleum products. This amounted to 34.1% of all seaborne

Table 13.2 Time charter equivalent rates, per day (2004–2015)

Class	*Cargo type*	*Route*	*2004*	*2005*	*2006*	*2010*	*2012*	*2014*	*2015*
VLCC	Crude	Persian Gulf – Japan	$95,250	$59,070	$51,550	$38,000	$20,000	$28,000	$57,000
Suezmax	Crude	West Africa – Caribbean or East Coast of North America	$64,800	$47,500	$46,000	$31,000	$18,000	$28,000	$4600
Aframax	Crude	Cross-Mediterranean	$43,915	$39,000	$31,750	$20,000	$15,000	$25,000	$37,000
All product carriers		Caribbean – East Coast of North America or Gulf of Mexico	$24,550	$25,240	$21,400	$11,000	$11,000	$12,000	$21,000

trade for the year 2005. Combining the amount carried with the distance it was carried, oil tankers moved the equivalent of 11,705 billion metric tonne-miles of oil. By comparison, in 1970, 1.44 billion metric tonnes of oil were shipped by tanker. This amounted to 34.1% of all seaborne trade for that year. In terms of amount carried and distance carried, oil tankers moved 6,487 billion metric tonne miles of oil in 1970. This represents a considerable increase of 5,218 billion barrels over just 35 years, suggesting worldwide consumption of oil has almost doubled. The United Nations maintains statistics about oil tanker productivity, which are stated in terms of metric tonnes carried per metric tonne of dwt, as well as metric tonne miles of carriage per metric tonne of dwt. In 2005, for each 1 tonne of dwt of oil tankers, 6.7 metric tonnes of cargo were carried. Similarly, each 1 tonne dwt of oil tankers was responsible for 32,400 metric tonne miles of carriage. The main loading ports are in Western Asia, Western Africa, North Africa, and the Caribbean, with 196.3, 196.3, 130.2, and 246.6 million metric tonnes of cargo loaded in these regions alone. The main discharge ports are in North America, Europe, and Japan with 537.7, 438.4, and 215.0 million metric tonnes of cargo discharged in these regions.

International law requires that every merchant ship be registered in a country. This country is called its Flag State. A ship's Flag State exercises regulatory control over the vessel and is required to inspect it regularly, certify the ship's equipment and crew, and issue safety and pollution prevention documents. In 2007, the US Central Intelligence Agency (CIA) statistics counted 4295 oil tankers of 1000 long tonnes dwt or greater worldwide. Panama was the world's largest Flag State for oil tankers, with 528 vessels in its registry. Six other Flag States had more than 200 registered oil tankers: Liberia (464), Singapore (355), China (252), Russia (250), the Marshall Islands (234), and the Bahamas (209). The Panamanian, Liberian, Marshallese, and Bahamian Flags are open registries and considered by the International Transport Workers' Federation to be Flags of Convenience. By comparison, the US and the UK had 59 and 27 registered oil tankers, respectively.

In 2005, the average age of oil tankers worldwide was 10 years. Of these, 31.6% were under 4 years old and 14.3% were over 20 years old. In 2005, 475 new oil tankers were built, accounting for 30.7 million dwt. The average size for these new tankers was 64,632 dwt. Nineteen of these were VLCC size, 19 were Suezmax, 51 were Aframax, and the remainder were of smaller designs. By comparison, 8.0 million dwt, 8.7 million dwt, and 20.8 million dwt worth of oil tanker capacity was built in 1980, 1990, and 2000, respectively. Ships are removed from the fleet through a process known as scrapping. Shipowners and buyers negotiate scrap prices based on factors such as the ship's empty weight (called the light tonne displacement or LDT) and prices in the scrap metal market. In 1998, almost 700 ships went through the scrapping process at shipbreakers in places such as Gadani (Pakistan), Alang (India), and Chittagong (Bangladesh). In 2004 and 2005, some 7.8 million dwt and 5.7 million dwt, respectively, were scrapped. Between 2000 and 2005, the capacity of oil tankers scrapped each year ranged from between 5.6 million dwt to 18.4 million dwt. In this same timeframe, tankers accounted for between 56.5% and 90.5% of the world's total scrapped ship tonnage. Within this period, the average age of scrapped oil tankers ranged from 26.9 to 31.5 years.

In 2005, the price for a new oil tanker in the 32,000–45,000 dwt, 80,000–105,000 dwt, and 250,000–280,000 dwt ranges were US$43 million, $58 million, and US$120 million, respectively. In 1985, these vessels would have cost US$18 million, US$22 million, and US$47 million, respectively. Oil tankers are often sold second hand. In 2005, 27.3 million dwt worth of used oil tankers were sold. Some representative prices for that year include US$42.5 million for a 40,000 dwt tanker, US$60.7 million for an 80,000–95,000 dwt tanker, US$73 million for a 130,000–150,000 dwt tanker, and US$116 million for a 250,000–280,000 dwt tanker. As an illustrative example, in 2006, Bonheur subsidiary First Olsen paid US$76.5 million for *Knock Sheen*, a 159,899 dwt tanker. Daily operating costs vary, but by current standards, a VLCC costs between US$10,000 and US$12,000 per day.

STRUCTURAL DESIGN AND ARCHITECTURE

Oil tankers have from eight to twelve tanks. Each tank is split into two or three independent compartments by fore-and-aft bulkheads. The tanks are numbered with tank one being the forwardmost. Individual compartments are referred to by the tank number and the athwartships position, such as 'one port', 'three starboard', or 'six centre'. The tanks are separated from each other by a cofferdam. A cofferdam is a small space left open between two bulkheads, to provide protection from heat, fire, or collision. Tankers have cofferdams forward and aft of the cargo tanks, and sometimes between individual tanks. A pumproom houses all the pumps connected to a tanker's cargo lines. Some larger tankers have two pumprooms. A pumproom spans the total breadth of the ship.

A major component of tanker architecture is the design of the hull or outer structure. A tanker with a single outer shell between the product and the ocean is said to be 'single-hulled'. Most newer tankers are 'double-hulled', with an extra space between the hull and the storage tanks. Hybrid designs such as 'double-bottom' and 'double-sided' hulls combine aspects of single and double-hull designs. MARPOL sets out the ambition to phase out single-hulled tankers by 2026, although the United Nations decided to phase out single-hulled tankers by 2010. In 1998, the Marine Board of the US National Academy of Sciences conducted a survey of industry experts regarding the advantages and disadvantages of double-hull designs. Some of the advantages of the double-hull design that were mentioned included the ease of ballasting in emergency situations, a reduction in the practice of using saltwater ballast in cargo tanks (therein decreasing corrosion), increased environmental protection, more efficient cargo discharge operations, more efficient tank washing, and improved protection in low-impact collisions and grounding. The same report also listed the following negative attributes to the double-hull design: higher build costs, greater operating expenses (e.g. higher canal and port tariffs), difficulties in ballast tank ventilation, the fact that ballast tanks need continuous monitoring and maintenance, increased transverse free surface, increased numbers of surfaces to maintain, increased risks of explosions in double-hull spaces in the absence of vapour detection systems, and increased cleaning and maintenance schedules. Despite these considerable drawbacks, double-hulled tankers are widely considered an improvement on single-hulled designs, especially when running aground on soft and rocky seabed. Although double-hull designs are superior in low energy casualties, preventing spillage in small casualties, in high energy casualties where both hulls are breached, oil can spill through the double-hull and into the sea. This is no less catastrophic than with single-hulled vessels, although spills from double-hulled tankers can be significantly higher than designs like the mid-deck tanker, the Coulombi Egg Tanker, and even pre-MARPOL tankers, as the latter has a lower oil column and reaches hydrostatic balance sooner.

An oil tanker's inert gas system is one of the most important parts of its design. Fuel oil itself is exceedingly difficult to ignite, but its hydrocarbon vapours are explosive when mixed with air in certain concentrations. The purpose of the inert gas system is to create an atmosphere inside tanks in which the hydrocarbon oil vapours cannot burn. As inert gas is introduced into a mixture of hydrocarbon vapours and air, it increases the lower flammable limit or lowest concentration at which the vapours can be ignited. At the same time, it decreases the upper flammable limit or highest concentration at which the vapours can ignite. When the total concentration of oxygen in the tank decreases to about 11%, the upper and lower flammable limits converge, and the flammable range disappears. Inert gas systems deliver air with an oxygen concentration of less than 5% by volume. As a tank is pumped out, it is filled with inert gas and kept in this safe state until the next cargo is loaded. The exception is in cases when the tank must be entered. Safely gas-freeing a tank is accomplished by purging hydrocarbon vapours with inert gas until the hydrocarbon concentration inside the tank is under about 1%. Thus, as air replaces the inert gas, the concentration cannot rise to the lower flammable limit and is safe.

CARGO OPERATIONS

Operations on board oil tankers are governed by an established body of best practices and a large body of international law. Cargo can be moved on or off an oil tanker in several ways. One method is for the ship to moor alongside a pier, and connect with cargo hoses or marine loading arms. Another method involves mooring to offshore buoys, such as a single point mooring, and making a cargo connection via underwater cargo hoses. A third method is ship-to-ship transfer, also known as lightering. In this method, two ships come alongside in an open sea with the oil transferred manifold to manifold via flexible hoses. Lightering is sometimes used where a loaded tanker is too large to enter a specific port.

Preparations for loading or unloading cargo

Prior to any transfer of cargo, the chief officer must develop a transfer plan detailing the specifics of the operation, such as how much cargo will be moved, which tanks will be cleaned, and how the ship's ballasting will change. The next step before a transfer is the pretransfer conference. The pretransfer conference covers issues such as what products will be moved, the order of movement, names and titles of key people, particulars of shipboard and shore equipment, critical states of the transfer, regulations in effect, emergency, and spill-containment procedures, watch and shift arrangements, and shutdown procedures. After the conference is complete, the person in charge on the ship and the person in charge of the shore installation go over a final inspection checklist. In the US, the checklist is called a Declaration of Inspection or DOI. Outside the US, the document is called the 'Ship / Shore Safety Checklist'. Items on the checklist include ensuring proper signals and signs are displayed, the secure mooring of the vessel, the choice of language for communications, the securing of all connections, confirmation that emergency equipment is in place and operable, and that no repair work is taking place.

Loading cargo

Loading an oil tanker consists primarily of pumping cargo into the ship's tanks. As oil enters the tank, the vapours inside the tank must somehow be expelled. Depending on local regulations, these vapours may be expelled into the atmosphere or discharged back to the pumping station by way of a vapour recovery line. It is common for the ship to decrease water ballast during the loading of cargo to maintain proper trim. Loading starts slowly at low pressure to ensure that the equipment is working correctly and that the connections are secure. Then, a steady pressure is achieved and held until the 'topping-off' phase when the tanks are full. Topping off is an extremely dangerous time in handling oil, and the procedure is managed with extreme care. Tank-gauging equipment is used to tell the person in charge how much space is left in the tank, and all tankers have at least two independent methods for tank-gauging. As the tanker reaches maximum capacity, crew members open and close valves to direct the flow of product and maintain close communication with the pumping facility to decrease and finally stop the flow of cargo.

Unloading cargo

The process of transferring oil off a tanker is like loading, except for some key differences. The first step in the operation is following the same pre-transfer procedures as used in loading. When the transfer begins, it is the ship's cargo pumps that are used to move the product ashore. As in loading, the transfer starts at a low pressure to ensure that the equipment is working correctly and that the connections are secure. Then, a steady pressure is achieved and held during the operation. While pumping, the tank levels are carefully monitored and key locations, such as the connection at the cargo manifold and the ship's pumproom, are constantly supervised. Under the direction of the person in charge, crew members open and close valves to direct the flow of cargo and maintain close communication with the receiving facility to decrease and finally stop the flow of cargo.

Tank cleaning

The tanks must be cleaned from time to time for several reasons. One reason is to change the type of product carried inside a tank. Also, when tanks are to be inspected or maintenance must be performed within a tank, it must be not only cleaned, but made gas-free. On most crude-oil tankers, a special crude oil washing (COW) system is a part of the cleaning process. The COW system circulates part of the cargo through the fixed tank-cleaning system to remove wax and asphaltic deposits. Tanks that carry less viscous cargoes are washed with water. Fixed and portable automated tank cleaning machines, which clean tanks with high-pressure water jets, are widely used. Some systems use rotating high-pressure water jets to spray hot water on all the internal surfaces of the tank. As the spraying takes place, the brown waste is pumped out of the tank. After a tank is cleaned, if it is going to be prepared for entry, it will be fully purged. Purging is accomplished by pumping inert gas into the tank until all traces of hydrocarbons have been sufficiently expelled. Next, the tank is gas freed, which is usually accomplished by blowing fresh air into the space with portable air-powered or water-powered air blowers. 'Gas freeing' brings the oxygen content of the tank up to 20.8%. The inert gas buffer between fuel and oxygen atmospheres ensures they are never capable of ignition. Specially trained personnel monitor the tank's atmosphere, often using hand-held gas indicators, which measure the percentage of hydrocarbons present. After a tank is gas-free, it may be further hand-cleaned in a manual process known as mucking. Mucking requires protocols for entry into confined spaces, protective clothing, designated safety observers, and the use of airline respirators.

SPECIAL USE OIL TANKERS

Some sub-types of oil tankers have evolved to meet specific military and economic needs. These sub-types include naval replenishment ships, oil-bulk-ore combination carriers, FSOs, and floating production storage and of FPSOs. Replenishment ships, known as oilers in the USA and fleet tankers in the UK and Commonwealth countries, are ships that provide oil products to naval vessels whilst underway. This process is called underway replenishment (USA) or replenishment at sea (RAS; UK). Replenishment at sea extends the length of time a naval vessel can remain at sea, as well as increases her effective range. Before ships could replenish at sea, naval vessels had to enter a port or anchor to take on fuel. In addition to fuel, replenishment ships may also deliver water, ammunition, rations, stores, and personnel. An ore-bulk-oil carrier, also known as a combination carrier or OBO, is a ship designed to be capable of carrying wet or dry bulk cargoes. This design was intended to provide flexibility in two ways. First, the OBO should be capable of switching between dry and wet bulk trades based on market conditions. Second, OBOs should be capable of carrying oil on one leg of a voyage and return carrying dry bulk, therein reducing the number of unprofitable ballast voyages the vessel would otherwise be forced to make. In practice, the flexibility that the OBO design allows has gone unused, as these ships tend to specialise in either the liquid or dry bulk trade. Also, these ships have endemic maintenance problems. On the one hand, due to less specialised design considerations, an OBO suffers more from wear and tear during dry cargo onload than normal bulk carriers. On the other hand, components of the liquid cargo system, from pumps to valves to piping, tend to develop problems when subjected to periods of disuse. These factors have contributed to a steady reduction in the number of OBO ships worldwide since they were introduced in the 1970s. The most famous OBO was the 180,000 dwt *MV Derbyshire*, which broke apart and sank in September 1980 during a Pacific typhoon whilst shipping iron ore from Canada to Japan. To this date, the *MV Derbyshire* is the single largest British ship to be lost at sea.

Floating storage and offloading units

Floating storage and offloading units are used worldwide by the offshore oil industry to receive oil from nearby platforms and store it until it can be offloaded onto oil tankers. A similar system, the FPSO, can process the product while it is on board. These floating units reduce oil production costs and offer mobility, large storage capacity, and production versatility. FPSOs and FSOs are often created out of old, stripped-down oil tankers, but can be made from new-built hulls. The first FPSO to come into use was a reconditioned tanker owned by Shell España in August 1977. These units are usually moored to the seabed through a spread mooring system. A turret-style mooring system can be used in areas prone to severe weather. This turret system allows the unit to rotate which helps minimise the effects of sea swell and wind. As we mentioned earlier, the largest oil tanker ever built, the *Knock Nevis*, was converted into an FPSO before being scrapped.

ENVIRONMENTAL IMPACT AND MARINE POLLUTION

Oil spills have devastating and long-lasting effects on the environment. Crude oil contains polycyclic aromatic hydrocarbons (PAHs), which are exceedingly difficult to clean up and last for years in the sediment and marine environment. Marine species constantly exposed to PAHs often exhibit developmental problems, susceptibility to disease, and abnormal reproductive cycles. By the sheer amount of oil carried, modern oil tankers provide an existential threat to the marine environment. As discussed above, a VLCC tanker can carry two million barrels (320,000 m^3) of crude oil. This is about eight times the amount spilt by the *Exxon Valdez* in March 1989. In this incident, the ship ran aground and dumped 10,800,000 US gallons (41,000 m^3) of crude oil into the ocean. Despite the efforts of scientists, environmentalists, and volunteers, over 400,000 seabirds, 1000 sea otters, and a catastrophic number of fish were killed. Considering the volume of oil carried by sea, however, tanker owners often argue that the industry's safety record is excellent, with only a tiny fraction of a percentage of oil cargoes carried being spilt. In fact, the International Association of Independent Tanker Owners (INTERTANKO) has observed that 'accidental oil spills this decade [2000–2010] have been at record low levels – one-third of the previous decade and one-tenth of the 1970s – at a time when oil transported has more than doubled since the mid-1980s'. Oil tankers are only one source of oil spills. According to the US Coast Guard, 35.7% of the volume of oil spilt in the US from 1991 to 2004 came from tank vessels (i.e. ships and barges), 27.6% from facilities and other non-vessels, 19.9% from non-tank vessels, 9.3% from pipelines, and 7.4% from mystery spills. Only 5% of the actual spills came from oil tankers, while 51.8% came from other kinds of vessels. Detailed statistics for 2004 show tank vessels responsible for less than 5% of the number of total spills but more than 60% of the volume. We can deduce therefore that whereas tanker spills are much rarer causes of oil spills, they tend to be several magnitudes more serious (Figure 13.3).

The (INTERTANKO) Pollution Federation has tracked 9,351 accidental spills that have occurred since 1974. According to this study, most spills result from routine operations such as loading cargo, discharging cargo, and taking on fuel oil. Ninety-one per cent of the operational oil spills are small, resulting in less than 7 metric tonnes per spill. On the other hand, spills resulting from accidents such as collisions, groundings, hull failures, and explosions are more numerous, with 84% of these involving losses of over seven hundred metric tonnes.

Figure 13.3 *MT Exxon Valdez*, 24 March 1989, Bligh Reef, Prince William Sound, Alaska.

Following the *Exxon Valdez* spill, the U.S. passed *the Oil Pollution Act of 1990* (OPA-90), which excluded single-hull tank vessels of 5000 metric tonnes or more from US waters from 2010 onwards, apart from those with a double bottom or double sides, which may be permitted to trade to the US through 2015, depending on their age. Following the sinkings of *Erika* (1999) and *Prestige* (2002), the EU has passed its own stringent anti-pollution packages (known as *Erika I, II*, and *III*), which also required all tankers entering EU waters to be double-hulled by 2010. The *Erika* packages are controversial because they introduced the new legal concept of 'serious negligence'. In addition, air pollution from engine operation and from cargo fires are other serious cause of environmental concern. Large ships often run off low-quality fuel oils, such as bunker oil, which is highly polluting and has been shown to be a health risk. Ship fires may result in the loss of the ship due to a lack of specialised firefighting gear and techniques, with fires often burning for days on end.

This completes Part II. In the next part, we will discuss passenger ships, including cargo liners, cruise ships, passenger ferries, cruise ferries, and ocean lines.

Part III

Passenger vessels

Chapter 14

Cargo liners

A cargo liner, also known as a passenger-cargo ship or passenger-cargoman, is a type of merchant ship which carries general cargo and often passengers. They became common just after the middle of the 19th century and eventually gave way to container ships and other more specialised carriers in the latter half of the 20th century. The main definition of a cargo liner is 'a vessel which operated a regular scheduled service on a fixed route between designated ports and carries many consignments of different commodities'. Cargo liners transport general freight, from raw materials to manufactures to merchandise. Many had cargo holds adapted to services, with refrigerator space for frozen meats or chilled fruit, tanks for liquid cargos such as plant oils, and lockers for valuables. Cargo liners typically carried passengers as well, usually in a single class. They differed from ocean liners, which focussed on the passenger trade, and from tramp steamers, which did not operate on regular schedules. Cargo liners sailed from port to port along routes and on schedules published in advance. The steam-powered cargo liner developed in the mid-19th century with the advancement of technology allowing bigger steamships to be built. As cargo liners were faster than tramp cargo ships, they were used for the transport of perishable and high-value goods, as well as providing passenger service. At first, they were mostly used in Europe and America as well as across the Atlantic Ocean between Europe and America. Longer routes, such as that to Oceania, remained in the hands of sailing ships a little bit longer, due to the inefficiency of the steamship of the time, until the late-1860s when the 1869 opening of the Suez Canal put sail ships at a distinct disadvantage.

The use and increased reliability of the compound steam engine gave greater fuel efficiency and opened these routes up to steamships. Alfred Holt pioneered the use of this type of engine in his fleet of steamships. By the last third of the 19th century, it was possible for a steamship to carry enough coal to travel 6,000 mi (9700 km) before needing to refuel. The opening of the Suez Canal in 1869 and the Panama Canal in 1914 also made the use of cargo liners more profitable and made possible regular scheduled overseas services. Cargo liners soon comprised 'the greater portion of the British merchant fleet', which at that time was the largest in the world.

DOI: 10.1201/9781003342366-18

With a focus on high-value freight, most cargo liners carried a limited number of passengers, most commonly 12, as British regulations required a doctor for ships with over twelve passengers on board. The decline of the cargo liner started in the early 1960s with the introduction of container ships. By the late 1970s, all cargo-liner services had been decommissioned and replaced with container liners. Several large container vessels still offer a small number of berths to paying passengers. Typically, a maximum of 12 passengers may be carried, as the ship would otherwise be legally required to carry a doctor on board. The recreational facilities are those used by the crew and may be limited to a lounge, a gym with exercise equipment, and a small swimming pool.

THE LAST CARGO LINER: *RMS ST. HELENA* (1989)

The *RMS St. Helena* was one of the last surviving cargo liners and served the British overseas territory of Saint Helena. She sailed between Cape Town, South Africa, and Saint Helena with regular shuttles continuing to Ascension Island. Some voyages also served Walvis Bay en route to and from, or occasionally instead of, Cape Town. She visited Portland, Dorset twice a year with normal calls in the Spanish ports of Vigo (northbound) and Tenerife (southbound) until 14 October 2011, when she set sail on her final voyage from Portland. On 10 February 2018, she departed for her last trip from St Helena to Cape Town. At the time of her retirement from St Helena service, she was one of only four ships in the world still carrying the status of Royal Mail Ship. Locals, including the local press, have usually called her the *RMS* rather than *St. Helena*, in order not to confuse her with the island itself. Formerly, Saint Helena Island was occasionally served by ships of the Union-Castle Line, which ran between the UK and South Africa. By the 1970s, the number of ships taking this route had declined significantly and the Union-Castle withdrew from the route completely at the end of 1977. As Saint Helena lacked an airfield, the British government had to purchase a ship to service the remote island and its dependencies from Cape Town. The British government purchased the part passenger, part cargo ship *Northland Prince* to fulfil the role of servicing Saint Helena, and after being refitted and renamed, this became the first *RMS St. Helena*. Originally built in 1963, this converted 3,150 tonne ship had room to carry 76 passengers and supplies. The ship was used by the Royal Navy during the Falklands War as a minesweeper support ship. By the 1980s, it was becoming apparent that the ship was too small for the island's needs, resulting in the purchase of the new *St. Helena*, which was built in 1989. As the island lacks a port suitable for large ships, the *RMS St. Helena* anchored near the island and loaded and unloaded cargo to and from lighters. The new *RMS St. Helena*, the last ship to be built in Aberdeen, Scotland, was launched by Hall, Russell & Company in 1989. The *RMS St. Helena* was a British registered Class 1 passenger/cargo ship and operated with a complement of 56 officers and crew. The vessel was equipped to carry a wide range of cargo, including liquids, to meet the needs of the population of Saint Helena. She also had berths for 155 passengers and associated facilities, including a swimming pool, shop, and lounges. She also had well-equipped medical facilities and an onboard doctor. In 2012, the ship's capacity was extended by the addition of 24 extra cabin berths, and a new gym. AW Ship Management offered a package deal where passengers could travel in one direction on the *RMS St. Helena* and in the other direction by taking Royal Air Force (RAF) flights to or from RAF Ascension Island and RAF Brize Norton in Brize Norton, England (Figure 14.1).

In November 1999, the *RMS St. Helena* broke down en route to the island and was forced into the French port of Brest to undergo repairs. Many people were left stranded on the island with no way in or out whilst the ship was being repaired. Panic-buying ensued as islanders became concerned about the non-delivery of vital supplies. This incident intensified calls for the island to be provided with an airport. On 25 August 2000, *RMS St. Helena* suffered a minor engine room fire while sailing from Cardiff to Tenerife on the first leg of her journey to the island. No one was injured and there was no considerable damage. In March and April 2017, several Cape Town – Saint Helena voyages were cancelled because of technical problems with the propellers, making the island isolated, as the airport was still not operational. In 2005, the British government announced plans to construct an

Figure 14.1 RMS St. Helena, James Bay, Island of St. Helena.

airport on Saint Helena, which would lead to the withdrawal from service of the *RMS St. Helena*. The airport was initially expected to be operational by 2010. However, it was not approved until October 2011, with work commencing in 2012. The estimated cost of the project was £240million and the airport was due to open in the first quarter of 2016. However, due to concerns about wind shear, on 26 April 2016 the Saint Helena Government announced an indefinite postponement to the opening of Saint Helena Airport. *RMS St. Helena* had been placed for disposal via London shipbrokers CW Kellock but was subsequently restored to service. The voyage originally intended as her final one began on 14 June 2016 from the UK and ended on 15 July in Cape Town, calling at Tenerife, Ascension Island, and Saint Helena. As part of its farewell voyage, Royal Mail organised a letter exchange with pupils from Cardiff and St. Helena. However, due to the postponed opening of the airport, the schedule of *RMS St. Helena* was extended as an interim measure. The ship was initially scheduled to run until July 2017, and then February 2018. After the opening of Saint Helena Airport to scheduled passenger flights on 14 October 2017, *RMS St. Helena* was finally withdrawn from service, and her last sailing from Saint Helena Island was on 10 February 2018.

Freight services for Saint Helena Island have since been taken over by the *MV Helena* cargo ship, which does carry a limited number of passengers, mail, and other express freight by the passenger aircraft. The first passenger on the *MV Helena* stated that unlike the *RMS Saint Helena*, the new ship, with a lower capacity, is strictly geared towards cargo, although some former RMS employees had become crew on the new ship. In April 2018, *RMS St. Helena* was purchased by MNG Maritime and entered service as a vessel-based armoury in the Gulf of Oman named *MNG Tahiti* to supply weaponry to ships travelling through the high-risk area of heightened pirate activity in the Indian Ocean. In October 2018, the vessel was resold to St Helena LLC, Jersey, and in 2019, the ship was refitted to function as a mobile hub for the race events of the Extreme E electric SUV racing series. Today, she is used to carry equipment, including cars, to various race locations worldwide.

In this brief chapter, we have looked at what was arguably the last remaining cargo liner to operate, the *RMS St. Helena*. Sadly, she was sold in 2019, effectively marking the end of an era in the history of the Merchant Navy. It is worth noting that some ship lines continue to offer limited passage on board though these are increasingly difficult to come by, and prohibitively expensive, and not particularly pleasant with short port calls and minimal interaction with the crew. In the next chapter, we will turn our attention to an entirely different category of passenger vessel, the cruise ship.

Chapter 15

Cruise ships

Cruise ships are large passenger ships used for vacationing. Unlike ocean liners, which are used for transport, they typically embark on round-trip voyages to various ports of call, where passengers may go on tours known as 'shore excursions'. On 'cruises to nowhere' or 'nowhere voyages', cruise ships make tw- to three-night round trips without visiting any ports of call. Modern cruise ships tend to have less hull strength, speed, and agility compared with ocean liners. However, they have added amenities to cater to water tourists, with recent vessels being described as 'balcony-laden floating condominiums'. As of December 2018, there were 314 cruise ships operating worldwide, with a combined capacity of 537,000 passengers. Cruising has become a major part of the global tourism industry, with an estimated market value of US$29.4 billion per year, and over 19 million passengers carried worldwide annually as of 2011. The industry's rapid growth saw nine or more newly built ships catering to a North American clientele added every year since 2001, as well as others servicing European clientele until the COVID-19 pandemic in 2020 saw the entire industry shutdown. As of 2022, the world's largest passenger ship is Royal Caribbean's *Wonder of the Seas*.

ORIGINS OF THE CRUISE SHIP

Italy, a traditional focus of the Grand Tour, offered an early cruise experience on the *Francesco I*, flying the flag of the Kingdom of the Two Sicilies. Built in 1831, the *Francesco I* sailed from Naples in early June 1833. Aristocrats, authorities, and royal princes from all over Europe boarded the cruise ship, which sailed in just over three months to Taormina, Catania, Syracuse, Malta, Corfu, Patras, Delphi, Zante, Athens, Smyrna, and Constantinople, delighting passengers with a combination of excursions and guided tours, dancing, card tables on deck, and parties on board. It was restricted to the aristocracy of Europe and was not a commercial undertaking. The British company P&O first introduced passenger cruising services in 1844, advertising sea tours to destinations such as Gibraltar, Malta, and Athens, with ships sailing from Southampton, England. The forerunner of modern

DOI: 10.1201/9781003342366-19

cruise holidays, these voyages were the first of their kind. To this day, P&O Cruises remains the world's oldest operating cruise line. The company later introduced round trips to destinations such as Alexandria in Egypt and Constantinople, Turkey. It underwent a period of rapid expansion in the latter half of the 19th century, commissioning larger and more luxurious ships to serve the steadily expanding market. Some of the most notable ships of this era included the *SS Ravenna* built in 1880, which became the first ship built with a total steel superstructure, and the *SS Valetta*, built in 1889, which was the first ship to use electric lights. The cruise of the German ship *Augusta Victoria* in the Mediterranean Sea and the Near East from 22 January to 22 March 1891, with 241 passengers [including Albert Ballin, the German shipping magnate and founder of the *Hamburg-Amerikanische Packetfahrt-Actien-Gesellschaft* (HAPAG) or Hamburg-America Line, and his wife], popularised cruises to a wider market. The first vessel built exclusively for luxury cruising was the *Prinzessin Victoria Luise*, which was designed by Albert Ballin. The ship was completed in 1900.

The practice of luxury cruising made steady inroads into the more established market for transatlantic crossings. In the competition for passengers, ocean liners – the *RMS Titanic* being the most famous example – added luxuries such as fine dining, luxury services, and staterooms with finer appointments. In the late 19th century, Albert Ballin was the first to send his transatlantic ships out on long southern cruises during the worst of the North Atlantic winter seasons. These proved surprisingly popular with other companies soon following suit. In 1897, three luxury liners, all European-owned, offered transportation between Europe and North America. In 1906, the number had increased to seven. British Inman Line owned the *City of Paris*, whereas Cunard Line had *Campania and Lucania*. The White Star Line owned the *Majestic* and *Teutonic*, and the French *Compagnie Générale Transatlantique* owned *La Lorraine* and *La Savoie* (Figure 15.1).

Figure 15.1 Marella Explorer at the Liverpool Cruise terminal, England.

From luxury liners to megaship cruising

With the advent of large passenger jet aircraft in the 1960s, intercontinental travellers started switching from ships to planes, sending the ocean liner trade into a terminal decline. Certain characteristics of older ocean liners made them unsuitable for cruising duties, such as high fuel consumption, deep draught preventing them from entering shallow ports, and cabins (often windowless) designed to maximise passenger numbers rather than comfort. Ocean liner services aimed at passengers ceased in 1986, with the notable exception of transatlantic crossings operated by the British Cunard Line, catering to a niche market of those who appreciated the several days at sea. To shift the focus of the market from passenger travel to cruising with entertainment value, Cunard Line pioneered the luxury cruise transatlantic service on board the ocean liner *Queen Elizabeth 2*. International celebrities were hired to perform cabaret acts onboard and the crossing was advertised as a vacation itself. *Queen Elizabeth 2* also inaugurated the concept of 'one-class cruising', where all passengers received the same quality berthing and facilities. This revitalised the market as the appeal of luxury cruising began to catch on, on both sides of the Atlantic. The 1970s television series *Love Boat* helped to popularise the concept of cruising as a romantic opportunity for couples. Another ship to make this transition was the *SS Norway*, originally the ocean liner *SS France*, which was converted to cruising duties as the Caribbean Sea's first 'super-ship'. Contemporary cruise ships built in the late 1980s and later, such as the *Sovereign class*, which broke the size record held for decades by the *SS Norway*, showed characteristics of size and strength once reserved for ocean liners. In fact, some have undertaken regular scheduled transatlantic crossings. The *Sovereign-class* ships were the first "megaships" to be built for the mass cruising market. They were also the first series of cruise ships to include a multi-story atrium with glass elevators. They had a single deck devoted entirely to cabins with private balconies instead of ocean view cabins. Other cruise lines soon launched ships with similar attributes, such as the *Fantasy class*, leading up to the Panamax-type *Vista class*, designed such that two-thirds of the ocean view staterooms have private verandas. As the veranda suites were particularly lucrative for cruise lines, something which was lacking in older ocean liners, recent cruise ships have been designed to maximise such amenities and have been described as 'balcony-laden floating condominiums'.

There have been nine or more new cruise ships added to the fleet every year since 2001, including the 11 members of the *Vista class*, with all being 100,000 metric-tonnes dwt or greater. The only comparable ocean liner to be completed in recent years has been Cunard Line's *Queen Mary 2*, which was launched in 2004. Following the retirement of her running mate *Queen Elizabeth 2* in November 2008, *Queen Mary 2* is the only ocean liner operating on transatlantic routes, though she also sees significant service on cruise routes. The *Queen Mary 2* was for a time the largest passenger ship

Figure 15.2 Color Magic.

before being surpassed by Royal Caribbean International's *Freedom class* vessels in 2006. The *Freedom class* ships were in turn overtaken by Royal Caribbean International's own *Oasis class* vessels, which entered service in 2009 and 2010. A distinctive feature of the *Oasis class* ships is the 'split open-atrium' structure, made possible by the hull's extraordinary width, with the 6-deck high 'Central Park' and 'Boardwalk' outdoor areas running down the middle of the ship, and verandas on all decks. In two short decades (1988–2009), the largest class cruise ships have grown one-third longer (from 268 to 360 m, or 879.3–1,181.1ft), almost doubled their beam (32.2–60.5 m, 105.8–198.6 ft), doubled the total passengers (2,744–5,400), and tripled in volume (73,000–225,000 dwt). Moreover, the 'megaships' went from a single deck having verandas to all decks with verandas (Figure 15.2).

CRUISE LINES

Operators of cruise ships are known as cruise lines, which are companies that sell cruises to the public. Cruise lines have a dual character; they are partly in the transportation business, and partly in the leisure entertainment business, a duality that carries down into the ships themselves, which have both a crew headed by the ship's captain and a hospitality staff headed by the equivalent of a hotel manager. Among cruise lines, some are direct descendants of the traditional passenger shipping lines (such as Cunard), while others were founded in the 1960s specifically for cruising. Historically, the cruise ship business has been volatile. The ships are large capital investments with high operating costs. A persistent decrease in bookings can put a company in financial jeopardy. Cruise lines have traditionally sold, renovated, or renamed their ships to keep up with travel trends. Cruise lines necessarily operate their ships constantly, though if the maintenance is unscheduled, this can result in thousands of dissatisfied customers. A wave of failures and consolidations in the 1990s led to many cruise lines being bought out by much larger holding companies, which continue to operate as 'brands' or subsidiaries of the holding company. Brands not only continue to be maintained partly because of the expectation of repeat customer loyalty, but also to offer various levels of quality and service. For instance, Carnival Corporation & Plc owns both Carnival Cruise Line, whose former image were vessels that had a reputation as 'party ships' for younger travellers, but have become large, modern, yet still profitable, and Holland America Line, whose ships cultivate an image of classic elegance. In 2004, Carnival merged Cunard's headquarters with that of Princess Cruises in Santa Clarita, California so that administrative, financial and technology services could be combined, ending Cunard's history of operating as a standalone company (subsidiary) regardless of parent ownership. Cunard did regain some operational independence in 2009 when its headquarters were moved to Carnival House in Southampton.

The customary practice in the cruise industry in listing cruise ship transfers and orders is to list the smaller operating company, not the larger holding corporation, as the recipient cruise line of the sale, transfer, or new order. In other words, Carnival Cruise Line and Holland America Line, for example, are the cruise lines from this common industry practice point of view, whereas Carnival Corporation & Plc and Royal Caribbean Cruises Ltd., for example, can be considered the holding corporations of the cruise lines. This industry practice of using the smaller operating company, not the larger holding corporation, is also followed in the list of cruise lines and in member-based reviews of cruise lines. Some cruise lines have specialties; for example, Saga Cruises only allows passengers over 50 years old on board their ships, whereas Star Clippers and formerly Windjammer Barefoot Cruises and Windstar Cruises only operate tall ships. Regent Seven Seas Cruises operates medium-sized vessels, a class smaller than the mega-ships operated by Carnival Corporation & Plc and Royal Caribbean and are designed such that all their suites have balconies. Several specialty lines offer 'expedition cruising' or only operate small ships, visiting obscure and distant destinations such as the Arctic and Antarctica, or the Galápagos Islands. The *John W. Brown*, which formerly operated as part of the US Merchant Marine during World War II before being converted to a museum ship, still gets underway several times a year for six-hour 'living history cruises' that take the ship through Baltimore Harbour, down the Patapsco River, and into the Chesapeake Bay. She is also the largest cruise ship operating under the American Flag on the US East Coast.

Currently, the three largest cruise line holding companies and operators in the world are Carnival Corporation & Plc, Royal Caribbean Cruises Ltd., and Norwegian Cruise Line Holdings. As an industry, the total number of cabins on all the world's cruise ships amount to less than 2% of the world's hotel rooms.

SHIPBOARD ORGANISATION

Cruise ships are organised much like floating hotels, with a complete hospitality staff in addition to the usual ship's crew. It is common for the most luxurious ships to have more crew and staff than passengers. Dining on all cruise ships is included in the cruise price. Traditionally, the ships' restaurants organise two dinner services per day, early dining and late dining, and passengers are allocated a set dining time for the entire cruise, although recently the trend is to allow diners to dine whenever they want. Having two dinner times allows the ship sufficient time and space to accommodate all its guests. That said, having two different dinner services can cause some conflicts with some of the ship's event programming (such as shows and performances) for the late diners, but this problem is usually fixed by having a shorter version of the event take place before the later dinner service. Cunard Line ships maintain the class tradition of ocean liners and have separate dining rooms for several types of suites, whereas Celebrity Cruises and Princess Cruises have a standard dining room and 'upgrade' specialty restaurants that require pre-booking and cover charges. Many cruises schedule one or more 'formal dining' nights, where the guests are expected to dress 'formally', however that is defined by the ship. The menu is typically more upscale than usual. Besides the dining room, modern cruise ships often contain one or more casual buffet-style eateries, which may be open 24 hours and with menus that vary throughout the day to provide meals ranging from breakfast to late-night snacks. In recent years, cruise lines have started to include a diverse range of ethnically themed restaurants on board each ship. Ships also feature numerous bars and nightclubs for passenger entertainment; most cruise lines do not include alcoholic beverages in their fares and passengers are expected to pay for drinks as they consume them. Most cruise lines also prohibit passengers from bringing aboard and consuming their own beverages, including alcohol, whilst on board. Alcohol purchased duty-free is sealed and returned to passengers when they disembark. There is often a central galley responsible for serving all major restaurants on board the ship, though specialty restaurants may have their own separate galleys. As with any vessel, adequate provisioning is crucial, especially on a cruise ship serving several thousand meals at each seating. For example, the Royal Princess requires a quasi 'military operation' to load and unload 3,600 passengers and eight metric tonnes of food at the beginning and end of each cruise.

Modern cruise ships typically have on board some or all of the following facilities: buffet restaurant, card room, casino (only open when the ship is at sea to avoid conflicts with local laws), childcare facilities, one or more cinemas, clubs, fitness centres, Jacuzzi and hot tubs, indoor and/or outdoor swimming pools with water slides, infirmary and morgue, karaoke, library, lounges, observation lounge, various retail outlets and duty-free shops (only open when the ship is at sea to avoid merchandising licensing and local taxes), spa, teen lounges, and theatre with West End and Broadway-style shows. Some ships have bowling alleys, ice skating rinks, rock climbing walls, sky-diving simulators, miniature golf courses, video arcades, zip-lines, surfing simulators, water slides, basketball courts, tennis courts, chain restaurants, ropes obstacle courses, and even roller coasters (Figure 15.3).

Crewing

The crew are usually hired on three-to-eleven-month contracts, which may then be renewed as mutually agreed, depending on the service ratings from passengers as well as the cyclical nature of the cruise line operator. Most staff work for 77 hours per week for 10 months continuously followed by two months of vacation. There are no paid vacations or pensions for service, non-management crew, depending on the level of the position and the type of the contract. Non-service and management crew members get paid

Figure 15.3 Cruise ship galley.

vacation, medical, retirement options, and can participate in the company's group insurance plan. The direct salary is low according to the North American standards, though restaurant staff have considerable earning potential from passenger tips. Crew members do not have any expenses while on board, as all food and accommodation, medical care, and transportation are included. Many unscrupulous crewing agencies often exploit the desperation of their employees. Living arrangements vary by cruise line, but mostly by shipboard position. In general, like most merchant ships, two crew members share a cabin with a shower, commode, and a desk with a television set, while senior officers are assigned single berth cabins. There is a set of facilities for the crew separate from that for passengers, such as mess rooms and bars, recreation rooms, prayer rooms/mosques, and fitness facilities, with some larger ships even having a crew deck with a swimming pool. All crew members are required to carry their certificates in accordance with the *International Convention on Standards of Training, Certification, and Watchkeeping for Seafarers, 1978* (as Amended), or complete the required training whilst on board. Crew members need to complete this certification prior to embarking as it is time-consuming and must be accomplished at the same time, they perform their daily work activities (Figure 15.4).

For the largest cruise operators, most 'hotel staff' are hired from less industrialised countries in Asia, Eastern Europe, the Caribbean, and Central America. While several cruise lines are headquartered in the USA, like most international shipping companies, ships are registered in countries such as

Figure 15.4 Carnival Vista off Oranjestad, Aruba.

the Netherlands, the UK, the Bahamas, and Panama. The International Labour Organisation's 2006 Maritime Labour Convention, also known as the 'Seafarers' Bill of Rights', provides comprehensive rights and protections for all crew members. The ILO sets rigorous standards regarding hours of work and rest, health, and safety, and living conditions for crew members, and requires governments to ensure that ships comply. For cruise routes around Hawaii, operators are required to register their ships in the USA and the crew is unionised, so these cruises are typically much more expensive than in the Caribbean or the Mediterranean.

BUSINESS MODEL

Most cruise lines since the 2000s have priced the cruising experience à la carte, as passenger spending aboard generates significantly more than ticket sales. The passenger's ticket includes the stateroom accommodation, room service, unlimited meals in the main dining room (or main restaurant) and buffet, access to shows, and the use of pool and gym facilities, while there is a daily gratuity charge to cover housekeeping and waiter service. However, there are extra charges for alcohol and soft drinks, official cruise photos, internet and wi-fi access, and specialty restaurants. Cruise lines earn significantly from selling onshore excursions offered by local contractors, often keeping 50% or more of what passengers spend for these tours. In addition, cruise ships earn significant commissions on sales from onshore stores that are promoted on board as 'preferred partners' (sometimes as much as 40% of gross sales). Facilitating this practice are modern cruise terminals with establishments of duty-free shops inside a perimeter accessible only by passengers and not by locals (similar in fact to most airports after Security). Ports of call have often oriented their own businesses and facilities towards meeting the needs of visiting cruise ships. In one case, Icy Strait Point in Alaska, the entire destination was created explicitly and solely for cruise ship visitors. Travel to and from the port of departure is usually the passengers' responsibility, although purchasing a transfer pass from the cruise line for the trip between the airport and cruise terminal will guarantee that the ship will not leave until the passenger is aboard. Similarly, if the passenger books a shore excursion with the cruise line and the tour runs late, the ship is obliged to remain until the passenger returns. Luxury cruise lines such as Regent Seven Seas Cruises and Crystal Cruises promote their fares as 'all-inclusive'. For example, the base fare on Regent Seven Seas ships includes most alcoholic beverages on board ship and most shore excursions in ports of call, as well as all gratuities that would normally be paid to hotel staff on the ship. The fare may also include a one-night hotel stay before boarding and the airfare to and from the cruise ship's origin and destination ports.

Cruise ship naming

Older cruise ships have often had multiple owners. It is usual for the transfer of ownership to entail a refitting and a name change. In fact, some older vessels have had a dozen or so identities. Many cruise lines have a common naming scheme they use for their ships. Some lines use their name as a prefix or suffix in the ship name such as the prefixes 'Carnival', 'AIDA', 'Disney', and 'Norwegian', and the suffix 'Princess'. Other lines use a unique word or phrase such as the prefix of 'Pacific' for P&O Cruises Australia or the suffixes 'of the Seas' for Royal Caribbean International and '-dam' for ships of the Holland America Line fleet. The addition of these prefixes and suffixes allows multiple cruise lines to use the same popular ship names while maintaining a unique identifier for each ship. It should be noted that this practice is not unique to the cruise industry, as many cargo vessel operators also use prefixes and suffixes in their ship naming conventions.

Cruise ship utilisation

Cruise ships and former liners often find use in applications other than those for which they were built. Due to slower speed and reduced seaworthiness, as well as being introduced after several major wars, cruise ships have also been used as troop transport vessels. By contrast, ocean liners were often seen as the pride of their country and used to rival liners of other nations and have been requisitioned during both World Wars and the Falklands War to transport soldiers and serve as hospital ships. During the 1992 Summer Olympics, 11 cruise ships docked at the Port of Barcelona for an average of 18 days, serving as floating hotels to help accommodate the large influx of visitors to the Games. They were available to sponsors and hosted 11,000 guests a day, making it the second-largest concentration of Olympic accommodation behind the Olympic Village. This hosting solution has been used since then in Games held in coastal cities, such as in Sydney in 2000, Athens in 2004, London in 2012, Sochi in 2014, Rio de Janeiro in 2016 and, had the Games not been postponed because of COVID-19, in Tokyo 2020 as well.

Cruise ships have been used to accommodate displaced persons during hurricanes. For example, on 1 September 2005, the US Federal Emergency Management Agency (FEMA) contracted three Carnival Cruise Lines vessels (*Carnival Fantasy*, the former *Carnival Holiday*, and the *Carnival Sensation*) to accommodate Hurricane Katrina evacuees. In 2017, cruise ships were used to help transport residents from some Caribbean islands destroyed by Hurricane Irma, as well as Puerto Rico residents displaced by Hurricane Maria. Cruise ships have also been used for evacuations, such as in 2010, in response to the shutdown of British airspace due to the eruption of Iceland's Eyjafjallajökull volcano. In this incident, the newly completed *Celebrity Eclipse* was used to rescue 2000 British tourists stranded in Spain as an act of goodwill by the owners. The ship departed from Southampton for Bilbao on 21 April and returned on 23 April. In Australia, a cruise ship was kept on standby in case inhabitants of Kangaroo Island required evacuation in 2020 after a series of bush fires burned rampant across the island.

Regional sectors

Most cruise ships sail the Caribbean or the Mediterranean. Others operate elsewhere in places like Alaska, the South Pacific, the Baltic Sea, and New England. A cruise ship that is moving from one of these regions to another will commonly operate a repositioning cruise while doing so. Expedition cruise lines, which usually operate small ships, visit certain more specialised destinations such as the Arctic and Antarctica, or the Galápagos Islands. The number of cruise tourists worldwide in 2005 was estimated at some 14 million. The main region for cruising was North America (with 70% of cruises), with the Caribbean islands as the most popular destinations. The second most popular region was continental Europe (13%), where the fastest-growing segment is cruising in the Baltic Sea. The most visited Baltic ports are Copenhagen (Denmark), St. Petersburg (Russia), Tallinn (Estonia), Stockholm (Sweden), and Helsinki (Finland). The seaport of St. Petersburg, the main Baltic port of call, received 426,500 passengers during the 2009 cruise season. Between 2008 and 2018, the Mediterranean cruise market was the fastest-growing cruise market worldwide with Italy winning the prime position as the number one destination for European cruises and the number one destination for the whole of the Mediterranean basin. The most visited ports in the Mediterranean Sea are Barcelona (Spain), Civitavecchia (Italy), Palma (Spain), and Venice (Italy). In 2013, the first Chinese company entered the global cruise market. China's first luxury cruise ship, *Henna*, made her maiden voyage from Sanya Phoenix Island International Port (China) in January 2013.

The Caribbean cruising industry is one of the largest in the world, responsible for over US$2 billion in direct revenue to the Caribbean islands each year. Over 45,000 people from the Caribbean are directly employed in the cruise industry. An estimated 17,457,600 cruise passengers visited the islands in the 2011–2012 cruise year (May 2011 to April 2012). Cruise lines operating in the Caribbean include Carnival Cruise Line, Celebrity Cruises, Crystal Cruises, Cunard, Disney Cruise Line, Holland America, Norwegian Cruise Line, P&O Cruises, Princess Cruises, Pullmantur Cruises, and Royal Caribbean International with the three largest cruise operators being Carnival Corporation & Plc, Royal Caribbean International, and Star Cruises/Norwegian Cruise Lines. There are also smaller cruise lines that cater to a more intimate feeling among their guests. Many American cruise lines to the Caribbean depart out of the Port of Miami, with nearly one-third of total cruises sailing out of Miami. Other cruise ships depart from Port Everglades (in Fort Lauderdale), Port Canaveral [approximately forty-five miles (72 km) east of Orlando], New York, Tampa, Galveston, New Orleans,

Figure 15.5 Costa Fortuna docked outside Dubrovnik's Old Town, Croatia.

Cape Liberty, Baltimore, Jacksonville, Charleston, Norfolk, Mobile, and San Juan, Puerto Rico. Some British cruise lines base their ships out of Barbados for the Caribbean season, operating direct charter flights out of the UK (Figure 15.5).

The busiest ports of call in the Caribbean for cruising in the 2013 year are summarised in Table 15.1.

2016 was the most recent year of CLIA (Cruise Lines International Association) studies conducted around the cruise industry specifically in the USA and more specifically Alaska. In 2016, Alaskan cruises generated five million passenger and crew visits, 20.3% of all passenger and crew visits in the USA. Cruise lines frequently bring passengers to Glacier Bay National Park, Ketchikan, Anchorage, Skagway, and the state's capital, Juneau. Visitor volume is represented by overall visitors entering Alaskan waters and/or airspace. Between October 2016 and September 2017, Alaska had about 2.2 million visitors, 49% of those were through the cruise industry. That 2.2 million was a 27% increase since 2009, and the volume overall has steadily increased. Visitors spend money when travelling, and this is measured in two distinct areas: the cruising companies themselves and the visitors. There are no current numbers for cruise-specific passenger spending ashore, but the overall visitor expenditure can be measured. Tours accounted for US$394 million (18%), gifts and souvenirs US$427 million (20%), food US$428 million (20%), transportation US$258 million (12%), lodging US$454 million (21%), and other US$217 million (10%).

Table 15.1 Busiest cruise destinations, 2013

Rank	*Destination*	*Pax arrivals*
1	Bahamas	4,709,236
2	Cozumel, Mexico	2,751,178
3	US Virgin Islands	1,998,579
4	Sint Maarten	1,779,384
5	Cayman Islands	1,375,872
6	Jamaica	1,288,184
7	Puerto Rico	1,176,343
8	Turks and Caicos Islands	778,920
9	Aruba	688,568
10	Belize	677,350
11	Haiti	643,634
12	Saint Kitts and Nevis	629,000
13	Curacao	610,186
14	Saint Lucia	594,118
15	Barbados	570,263
16	Antigua and Barbuda	533,993
17	Dominican Republic	423,910
18	British Virgin Islands	367,362
19	Bermuda	320,090
20	Dominica	230,588
21	Grenada	197,311
22	Martinique	103,770
23	Bonaire	96,818
24	Saint Vincent and the Grenadines	82,974

The second main area of economic growth comes from what the cruising companies and their crews spend themselves. Cruise liners spend around US$297 million on the items that come in their packages on board and ashore as parts of group tours: things like stagecoach rides and boat tours on smaller vessels throughout their ports of call. This money is paid to the service providers by the cruise line company. Cruise liner crew are also a revenue generator, with 27,000 crew members visiting Alaska in 2017 alone, generating about US$22 million. 2017 was also a good year for job generation within Alaska: 43,300 jobs were created, bringing in US$1.5 billion in labour costs, and a total income of US$4.5 billion was generated. These jobs were scattered across all Alaska. Southeast Alaska had 11,925 jobs (US$455 million labour income), the Southwest 1,800 jobs (US$50 million labour income), South Central 20,700 jobs (US$761 million labour income), Interior 8,500 jobs (US$276 million labour income), Far North 375 jobs (US$13 million labour income).

SHIPYARDS

The construction market for cruise ships is dominated by three European companies and one Asian company:

- Chantiers de l'Atlantique of France;
- Fincantieri of Italy with:
  - o Ancona shipyards (located at Ancona);
  - o Marghera shipyards (located at Marghera, Venice);
  - o Monfalcone shipyards (located at Monfalcone, Gorizia);
  - o Sestri Ponente shipyards (located at Genoa);
- VARD Braila shipyards (located at Braila);
- VARD Søviknes Shipyard (located in Norway);
- VARD Tulcea shipyards (located at Tulcea);
- Meyer Werft of Germany with two shipyards:
  - o Meyer Turku at Perno shipyard in Turku, Finland;
  - o Meyer Werft of Germany;
- Mitsubishi Heavy Industries of Japan.

A considerable number of cruise ships have been built by other shipyards, but no other individual yard has reached the volume of hulls achieved by the abovementioned five companies.

SAFETY AND SECURITY

Piracy and terrorism

As most of the passengers on a cruise are affluent and have considerable ransom potential, not to mention a considerable amount of cash and jewellery on board (e.g. in casinos and shops), there have been several high-profile pirate attacks on cruise ships, such as *on Seabourn Spirit* (2005) and *MSC Melody* (2009). As a result, cruise ships have implemented various security measures. While most merchant shipping firms have avoided arming crew or security guards for reasons of safety, liability, and conformity with the laws of the countries where they dock, cruise ships have small arms (usually semi-automatic pistols) stored in a safe accessible only by the captain who distributes them to authorised personnel such as security or the Master-at-Arms. The ship's high-pressure fire hoses can be used to keep boarders at bay, and often the vessel itself can be manoeuvered to ram pirate craft. Recent technology to be developed to deter pirates has been the LRAD or sonic cannon, which was used in the successful defence of the *Seabourn Spirit*. A related risk is that of terrorism, the most notable incident being that of the hijacking of *Achille Lauro*, an Italian cruise ship, in 1985.

Crime on board

Passengers entering the cruise ship are screened by metal detectors. Explosive detection machines include x-ray machines and explosives trace-detection portal machines (aka 'puffer machines') to prevent weapons, drugs, and other contraband on board. Port and ship security has been tightened since 11 September 2001, such that these measures are like airport security. In addition to security checkpoints, passengers are often given a ship-specific identification card, which must be shown to get on or off the ship. This prevents people boarding who are not entitled to do so and ensures that the ship's crew are aware of who is on the ship. The cruise ship's ID cards are also used as the passenger's room key. Most cruise ships make extensive use of CCTV throughout the vessel. In 2010, the US Congress passed the *Cruise Vessel Security and Safety Act after numerous incidents* involving sexual violence, passenger disappearances, physical assaults, and other serious crimes. In a report issued by the US Congress, it was claimed that

> passengers on cruise vessels have an inadequate appreciation of their potential vulnerability to crime while on ocean voyages, and those who may be victimised lack the information they need to understand their legal rights or to know whom to contact for help in the immediate aftermath of the crime.

Congress went on to state that 'both passengers and crew committed crimes and that data on the problem was lacking because cruise lines did not make it publicly available, multiple countries engaged in investigating incidents on international waters, and crime scenes could not be secured quickly by police'. The report recommended the owners of cruise vessels install acoustic hailing and warning devices capable of working at a distance, install more security cameras around their ships, install peep holes in passenger cabin doors, and limit access to passenger cabins to select staff at specific times.

After investigating the death of Dianne Brimble in 2002, a coroner in Australia recommended that Australian Federal Police officers travel on ships to ensure a quick response to crime, scanners and drug detection dogs check passengers and crew at Australian ports, an end to overlaps between jurisdictions, and Flags of ships be disregarded for nations unable to investigate incidents thoroughly and competently. The lobby group International Cruise Victims Association, based in Arizona, North America, also pushes for more regulation of the cruise industry and supports victims of crimes committed on cruise ships.

Overboard drownings

Passengers and crew sometimes drown after going overboard in what are called 'man-overboard' incidents (MOBs). Since 2000, more than 300 people have fallen off cruise ships or large ferries, which is an average of about 1.5 people each month. Of those, only about 17–25% of people were rescued. Critics of the industry blame alcohol promotion for many passenger deaths, and poor labour conditions for crew suicides. They also point to underinvestment in the latest MOB sensors, a lack of regulation and consumer protection, and a lack of onboard counselling services for crew members. The industry blames the irresponsible behaviour of passengers and says overboard sensors are unreliable and generate false alarms. One authority on overboard drownings, the maritime lawyer, James Walker, estimates about half of all disappearances at sea involve some factor of foul play, and that a lack of police authority on international waters allows sexual predators to go unpunished. According to the Washington Post, a recent study by economic consultant G.P. Wild – commissioned by the cruise industry's trade group and released in March 2019 – argued that cruises are getting safer over time. The study claims that, even as capacity has increased 55% between 2009 and 2018, the number of overall 'operational incidents' declined 37% and the rate of man overboard cases dropped 35%.

Stability

Modern cruise ships are tall but remain stable due to their low centre of mass. This is due to large open spaces and the extensive use of aluminium, high-strength steel and other lightweight materials in the upper parts, and the fact that the heaviest components, the engines, propellers, fuel tanks, etc., are located at the bottom of the hull. Thus, even though modern cruise ships may appear tall, proper weight distribution ensures that they are not top-heavy. Furthermore, large cruise ships tend to be very wide, which increases their initial stability by increasing the metacentric height. Although most passenger ships use stabilisers to reduce rolling in heavy weather, they are only used for crew and passenger comfort and do not contribute to the overall intact stability of the vessel. The ships must fulfil all stability requirements even with the stabiliser fins retracted.

Health concerns

Norovirus

Norovirus is a virus that commonly causes gastroenteritis in developed countries and is also a cause of gastroenteritis on cruise ships. It is typically transmitted from person to person. Symptoms usually last between one and three days and resolve without treatment or long-term consequences. The incubation period of the virus averages about 24 hours. Norovirus outbreaks are often perceived to be associated with cruise ships. According to the US Centres for Disease Control and Prevention (CDC), the factors that cause norovirus associated with cruise ships include the closer tracking and faster reporting of illnesses compared with those on land; the close living quarters that increases the amount of interpersonal contact; as well as the turnover of passengers that may bring the viruses on board. However, the estimated likelihood of contracting gastroenteritis from any cause on an average seven-day cruise is estimated to be less than 1%. In 2009, during which more than 13 million people participated in cruises worldwide, there were nine cruise ship-related reports of norovirus outbreaks. Outbreak investigations by the CDC have shown that transmission among cruise ship passengers is primarily person-to-person; potable water supplies have not been implicated. In a study published in the Journal of the American Medical Association, the CDC reported that; perceptions that cruise ships can be luxury breeding grounds for acute gastroenteritis outbreaks do not hold water. A recent CDC report showed that from 2008 to 2014, only 0.18% of more than seventy-three million cruise passengers and 0.15% of some twenty-eight million crew members reported symptoms of the illness'.

Ships docked in port undergo surprise health inspections. In 2009, ships that underwent unannounced inspections by the CDC received an average CDC Vessel Sanitation Programme score of approximately 97 out of a total possible 100 points. The minimum passing inspection score is 85 out of 100. Collaboration with the CDC's Vessel Sanitation Programme and the development of Outbreak Prevention and Response Plans has been credited with decreasing the incidence of norovirus outbreaks on cruise ships.

Legionnaires' disease

Other pathogens which can colonise in pools and spas including those on cruise ships include Legionella, the bacterium which causes Legionnaires' disease. Legionella, and the most virulent strain, Legionella pneumophila serogroup 1, can cause infections when inhaled as an aerosol or aspirated. Individuals who are immunocompromised and those with pre-existing chronic respiratory and cardiac disease are more susceptible. Legionnaires has been infrequently associated with cruise ships. The Cruise Industry Vessel Sanitation Programme has specific public health requirements to control and prevent Legionella.

Enterotoxigenic Escherichia coli

Enterotoxigenic *Escherichia coli* is a form of *E. coli* and the leading bacterial cause of diarrhoea in the developing world, as well as the most common cause of diarrhoea for travellers to those areas. Between 2008 and 2014, there was at least one reported incident each year of *E. coli* on international cruise ships reported to the Vessel Sanitation Programme, though there were no reported cases in 2015. Causes of *E. coli* infection include the consumption of contaminated food or water contaminated by human waste.

Coronavirus disease 2019 (COVID-19)

News outlets reported several cases and suspected cases of coronavirus disease 2019 (COVID-19) associated with cruise ships in early 2020. Authorities variously turned away ships or quarantined them; cruise operators cancelled some port visits and suspended global cruise operations. It is strongly believed that passengers on board cruise ships played a role in spreading the disease in some countries, though the scientific evidence supporting this aspersion is limited. After disembarking passengers, many cruise ships remained docked or at sea near ports with crew members still aboard, in some cases unpaid though with free room and board. Crew members are citizens of many different countries, aside from those where the ship is registered, based, or temporarily stopped. Especially with pandemic-related travel restrictions and concerns about onboard outbreaks, crew members could not simply disembark and fly home. After some negotiations, most cruise companies made special arrangements to allow crew members to take charter flights back to their home countries. In some cases, the ships themselves were used to transport crew members to home countries directly. According to the Miami Herald, Royal Caribbean International was slower than other companies to make these arrangements; in the meantime, two crew members died after jumping overboard, and several crew members on the *Navigator of the Seas* went on hunger strike demanding to be released. In total, there were eight suspected suicides or suicide attempts in May and June 2020. The government of Grenada complained that cruise companies that promised to pay for onshore quarantines failed to do so, and some were concerned about crew members being released into Haiti from Royal Caribbean ships without testing or onshore quarantine.

ENVIRONMENTAL IMPACT

Cruise ships generate several waste streams that can result in discharges to the marine environment, including sewage, grey water, hazardous wastes, oily bilge water, ballast water, and solid waste. They also emit air pollutants to the air and water. These wastes, if not properly treated and disposed of, can be a significant source of pathogens, nutrients, and toxic substances with the potential to threaten human health and damage aquatic life. Most cruise ships run (primarily) on heavy fuel oil (HFO) or bunker fuel, which, because of its high sulphur content, results in sulphur dioxide emissions worse than those of equivalent road traffic. The international MARPOL IV-14 agreement for sulphur emission control areas require less than 0.10% sulphur in marine fuel, contrasting with HFO. Cruise ships may use 60% of the fuel energy for propulsion, and 40% for hotel functions, but loads and distribution depend on sea conditions. Some cruise lines, such as Cunard, have taken steps to reduce environmental impact by refraining from discharges at sea (the *Queen Mary 2*, for example, has a zero-discharge policy) and reducing their carbon dioxide output each year. Cruise ships require electrical power, normally provided by diesel generators, although an increasing number of new ships are fuelled by LNG. When docked, ships must run their generators continuously to power the on-board facilities, unless they can use onshore power, where available. Some cruise ships already support the use of shore power, while others are being adapted to do so.

Chapter 16

Cruise ferry

A cruise ferry is a ship that combines the features of a cruise ship with a ROPAX ferry. Many passengers travel with the ships for the cruise experience, staying only a few hours at the destination port or not leaving the ship at all, while others use the ships as a means of transportation. Cruise ferry traffic is concentrated in the seas of Northern Europe, especially the Baltic Sea and the North Sea. However, similar ships traffic across the English Channel as well as the Irish Sea, Mediterranean, and even on the North Atlantic coast around Canada. Cruise ferries also operate from India, China, and between Australia and Tasmania. In the northern Baltic Sea, two major rival companies, Viking Line and Silja Line, have for decades competed on the routes between Turku and Helsinki in Finland and Sweden's capital Stockholm. Since the 1990s, Tallink has also risen as a major company in the area, culminating with the acquisition of Silja Line in 2006. The Baltic Sea is crossed by several cruise ferry lines, the largest being Viking Line, Silja Line, Tallink, St. Peter Line, and Eckerö Line.

CRUISE FERRY SECTORS

Eastern Baltic Sea

Upon departure from or arrival in Helsinki, Baltic Sea cruise ferries pass the island fortress of Suomenlinna. Tallink and Viking Line operate competing cruise ferries on the routes between Stockholm – Turku and Stockholm – Helsinki, calling in Åland (Mariehamn or Långnäs). Additionally, Tallink sails between Stockholm – Mariehamn – Tallinn and Stockholm – Riga. Tallink, Viking Line, and Eckerö Line compete on the Helsinki – Tallinn route, which is also the busiest route in the Baltic Sea, and was travelled by over 7.5 million people in 2018. Baltic routes are mostly served by new ships purpose-built for the passage. Older cruise ferries from the Baltic serve as ferries on other seas, or in some cases, as cruise ships. Viking Line and Eckerölinjen also operate short routes from Sweden to Åland, sailing on Kapellskär – Mariehamn and Grisslehamn – Berghamn. Birka Line, owned

DOI: 10.1201/9781003342366-20

by Eckerö, also operates short cruises out of Stockholm. *GTS Finnjet* (1977) is the first cruise ferry, as she was the first ferry to offer cruise-ship quality services and accommodations. The first generation of cruise ferries operating from Finland to Sweden was highly influenced by *Finnjet's* interior and exterior designs. After the fall of the Soviet Union, the route connecting Helsinki to Tallinn became highly lucrative, which led to Estonia-based company Tallink to grow and rival the two long-established companies operating in the Eastern Baltic, Viking Line, and Silja Line. Tallink purchased Silja Line in 2006. The size of Baltic cruise ferries is limited by various narrow passages in the Stockholm, Ålandian, and Turku archipelagos, meaning that ships with a length overall more than 200 m (656 ft) cannot service these routes. The single narrowest point is Kustaanmiekka strait outside Helsinki, although ships making port at the city's west harbour do not have to pass through the strait. Viking and Silja Line have wished to keep their terminals in the South Harbour, however, as it is located within the proximity of Helsinki city centre. The longest ships to maintain a scheduled service through the Kustaanmiekka strait were *MS Finnstar* and her sisters with a length overall of 219 m (718 ft). The longest ship to have ever navigated through the narrows past Suomenlinna sea fortress was *MS Oriana* (260 m, 853 ft), but that was only possible due to extremely clear weather conditions (Figure 16.1).

To allow for duty-free sales on routes between Finland and Sweden, Baltic Sea cruise ferries stop in Åland on the way. This often happens in the middle of the night, with the ships staying at the Åland Islands for less than an hour. Passengers board and depart the ships via walking tunnels connecting the ship directly to the terminal building. The expansion of the EU has limited the growth of the industry, as duty-free sales on intra-EU routes are no longer possible. However, as Åland is outside the EU customs zone,

Figure 16.1 Scandlines' *Prins Richard* on the Puttgarden-Rødby 'Vogelfluglinie' from Puttgarden, Germany to Rødby, Denmark.

duty-free sales are still possible on routes making a stop at Mariehamn or other harbours on the islands. Another popular destination is Estonia with its lower taxes on alcohol. The ferries have attracted criticism for their low prices of alcoholic beverages, which has been argued that it encourages passengers to become drunk and act irresponsibly. Due to the cheap price of the cruises and the availability of duty-free alcohol [which makes it cheaper than on 'land' (both Finland and Sweden have a strict taxation regime for alcohol], it is common for groups to hold parties involving the consumption of vast amounts of alcohol. Many Finns also buy snus[1] from ferries, as its sale is illegal in Finland due to the EU regulations.

Silja Line

Silja Line is a Finnish cruise ferry brand operated by the Estonian ferry company AS Tallink Grupp, for car, cargo, and passenger traffic between Finland and Sweden. The former company Silja Oy, today Tallink Silja Oy, is a subsidiary of the Tallink Grupp, handling marketing and sales for Tallink and Silja Line brands in Finland as well as managing Tallink Silja's shipboard personnel. Another subsidiary, Tallink Silja AB, manages marketing and sales in Sweden. Strategical corporate management is performed by the Tallink Grupp, which also owns the company's fleet of ships. As of 2018, four ships service two routes under the Silja Line brand, transporting about three million passengers and 200,000 cars every year. The Silja Line ships have a market share of around 50% on the two routes served. The history of Silja Line can be traced back to 1904 when the two Finnish shipping companies, Finland Steamship Company (Finska Ångfartygs Aktiebolaget, FÅA for short) and Steamship Company Bore, started collaborating on the Finland–Sweden traffic. The initial collaboration agreement was terminated in 1909 but was re-established in 1910. After the end of World War I in 1918, a new agreement was made that also included the Swedish Rederi AB Svea. Originally, the collaboration agreement applied only on services between Turku and Stockholm but was later adopted to the Helsinki–Stockholm route in 1928. As a precursor to the policies later adopted by Silja Line, each of the three companies ordered a near-identical ship for the Helsinki–Stockholm service to coincide with the 1952 Summer Olympics, which were held in Helsinki. In the end, only Finland SS Company's *SS Aallotar* was ready in time for the Olympics. At the same time, the city of Helsinki constructed the Olympia Terminal in Helsinki's South Harbour, which Silja Line ships still use today (Figure 16.2).

Realising that car-passenger ferries would be the dominant traffic form of the future, the three collaborating companies decided to form a daughter company, Oy Siljavarustamo/Siljarederiet AB. The new company started operations with used ships, which were not particularly well-fitted for the role they were meant for, but in 1961, Silja took delivery of the new *MS Skandia*, the first purpose-built car-passenger ferry to operate in the

Figure 16.2 St Peter Line's cruise ferry *MS Princess Maria*, outside Helsinki, Finland.

northern Baltic Sea. The *MS Skandia's* sister ship, the *MS Nordia* followed the next year culminating with the launch of the *MS Fennia* in 1966. Two more ships based on the Skandia design, *MS Botnia* and *MS Floria*, were delivered in 1967 and 1970, respectively. Despite the establishment of Silja, FÅA, Bore, and Svea also continued to operate on the same routes with their own ships. This led to a complex situation where four different companies were marketed as one entity. In Finland, they went by the name *Ruotsinlaivat* ('Sweden's Ships' or 'Ships to Sweden'), whereas in Sweden, the preferred terms were *Det Samseglande* (roughly 'the ones that sail together'), *Finlandsbåten* ('Finland's Ships'), or *Sverigebåten* ('Sweden Ships'). In both countries, the names of all four companies were usually displayed alongside the group identity. In 1967, three of Silja's rival companies had formed a joint marketing and coordination company, Viking Line, which was to become Silja Line's main rival for the next two decades. FÅA, Bore, and Svea soon realised that a similar arrangement would be preferable to their current fragmented image, and in 1970, a substantial change was conducted within the organisations: Silja Line was established as a joint marketing and coordination company between FÅA, Bore, and Svea, and the ships of Siljavarustamo were divided between the three companies. All Silja Line ships were painted in the same colour scheme, with a white hull and superstructure, with Silja Line and the seal's head logo on the side in dark blue. Each company retained their own funnel colours so that it was easy to distinguish which ship belonged to which company even from a distance:

Svea's funnels were white with a large black 'S,' FÅA's were black with two white bands, and Bore's were yellow with a blue and white cross. Already before the reorganisation, Silja had ordered two new ships from Dubigeon-Normandie SA of Nantes to begin year-round services between Helsinki and Stockholm. Until then, the route was offered during the summer season only. In 1972, these vessels were delivered to FÅA and Svea as *MS Aallotar* and *MS Svea Regina*, respectively. Passenger numbers on the Helsinki route grew fast, and already in 1973, it was decided that the three companies would each order a ship of identical design from the same shipyard to replace the current Helsinki – Stockholm ships. These were delivered in 1975, with the first being *MS Svea Corona*, followed by *MS Wellamo*, and the last, *MS Bore Star*. However, winter passenger numbers were insufficient for three ships, and as a result, *MS Bore Star* was chartered to Finnlines during the winter season of 1975–1976 and 1976–1977. In 1976, Finland SS Co changed its name to Effoa (the Finnish phonetic spelling of FÅA). During the latter part of the 1970s Effoa's old ferries, *MS Ilmatar* and *MS Regina* cruised the Baltic, Norwegian fjords, and the Atlantic (from Málaga) under the marketing name Silja Cruises.

In 1979, Svea and Effoa again decided to order new ships for the Helsinki–Stockholm route, which would be the largest ferries of their time. Bore, however, decided not to participate in building new ships, and in 1980, opted to bow out of passenger traffic altogether (Bore Line still exists as a freight-carrying company). Their two ships were sold to Effoa and their shares of Silja Line were split between the two other companies. In Finland, and later in Sweden, a large maritime strike in the spring of 1980 stopped ferry traffic completely and prompted Effoa to terminate the Silja Cruises service. Despite the difficulties, Silja's first real cruise ferries, *MS Finlandia* and *MS Silvia Regina*, entered service in 1981, which led to a 45% increase in passenger numbers. Later in the same year, Johnson Line purchased Rederi AB Svea, and the former Svea ships received Johnson Line's blue and yellow colours. The positive experiences with the new Helsinki ships prompted Effoa and Johnson Line to order a further two ships built on a similar principle for traffic on the Turku (Finland) to Stockholm route, which were delivered in 1985 and 1986 as *MS Svea* and *MS Wellamo*. Although similar in proportions and interior layout, the new ships sported an attractive streamlined superstructure instead of the box-like superstructure of the *MS Finlandia* and *Silvia Regina*. In 1987, Effoa purchased *the GTS Finnjet*, and from the beginning of 1987, the prestigious but unprofitable *Queen of the Baltic Sea* joined Silja Line's fleet. Later in the same year, Effoa and Johnson Line jointly purchased Rederi Ab Sally, one of the owners of the rival Viking Line. The other Viking Line partners forced the new owners to sell their share in Viking, but Effoa and Johnson Line retained Vaasanlaivat/Vasabåtarna, Sally Cruises, Sally Ferries UK, and Commodore Cruise Line. Although the purchase of Sally had no effect on Silja Line's traffic for the time being, it proved to be important later. Finally, 1987 saw another order of new ships for the Helsinki–Stockholm route, which would again be the largest ferries built at that time, which were eventually named *MS Silja Serenade* and *MS Silja Symphony*. Not revealed at the time, the new ships had a 140 m (459 ft) promenade street running along the centre of the ship, a feature never seen before on a vessel, but by the first decade of the 21st century, commonly found on Royal Caribbean International's and Color Line's newer ships.

In late 1989, Wärtsilä Marine, the shipyard building Silja's new cruise ferries, went bankrupt, which led to the ships being delivered later than had been planned. To ensure the delivery of their ferries, both Effoa and Johnson Line purchased a part of the new Masa-Yards established to continue shipbuilding in Wärtsilä's former shipyards. In 1990, Effoa and Johnson Line merged to form EffJohn. As a result, the seal's head logo replaced the colours of each individual owner company on the funnel. In November, the new *MS Silja Serenade* made its maiden voyage from Helsinki to Stockholm, approximately seven months after the originally planned delivery date. *MS Silja Symphony* was delivered the following year. Although popular and sporting a successful design, the new ships proved expensive. This expense, coupled with economic depression in the early 1990s, forced EffJohn to cut costs, which resulted in Wasa Line and Sally Cruises being merged into Silja Line in 1992. Also in 1992, *MS Svea* and *MS Wellamo* were modernised and renamed *Silja Karneval* and *Silja Festival*, respectively. In January 1993, it was reported that Silja Line had chartered the *MS Europa*, a ship under construction for Rederi AB Slite, one of the owners of Viking Line. Because of financial troubles, Slite could not pay for their new ship, and the shipyard decided to charter it to Silja instead. Later in the same year, Silja joined forces with Euroway on their Malmö (Sweden) – Travemünde (Germany) – Lübeck (Germany) route. The route proved unprofitable and was terminated in the spring of 1994. In September 1994, the worst peace-time maritime disaster on the Baltic Sea occurred with the sinking of the *MS Estonia*. *Silja Europa*, *Silja Symphony*, and *Finnjet* all assisted in searching for survivors from the disaster. *Silja Festival* was berthed opposite the *MS Estonia* in Tallinn the day before the sinking but arrived in Helsinki after the *MS Estonia* sank, and so did not come to her aid. The Estonia sinking led to passenger numbers dropping, which did not help Silja's precarious financial situation. Although the company had achieved its ambition of becoming the largest operator of cruise ferries on the Baltic Sea, having finally overtaken Viking Line in 1993, the company was not doing financially well. In 1995, Effjohn changed its name to Silja Oy Ab, before changing it again three years later to Neptun Maritime.

Viking Line

Viking Line Abp is a Finnish shipping company that operates a fleet of ferries and cruise ferries between Finland, the Åland Islands, Sweden, and Estonia.

The history of Viking Line can be traced back to 1959, when a group of seamen and businessmen from the Åland Islands province in Finland formed Rederi Ab Vikinglinjen, purchasing a steam-powered car ferry, the *SS Dinard* from the UK. The vessel was renamed the *SS Viking* and began service on the route between Korpo (Finland), Mariehamn (Åland), and Gräddö (Sweden). In the same year, the Gotland-based Rederi AB Slite began a service between Simpnäs (Sweden) and Mariehamn. In 1962, a disagreement caused a group of people to leave Rederi Ab Vikinglinjen and form a new company, Rederi Ab Ålandsfärjan, which began a service linking Gräddö and Mariehamn the following year. Soon the three companies, all competing for passengers between the Åland Islands and Sweden, realised that they eventually all stood to lose from the mutual competition. In 1965, Vikinglinjen and Slite began collaborating, and by the end of July 1966, Viking Line was established as a marketing company for all three companies. At this time, Rederi Ab Vikinglinjen changed its name to Rederi Ab Solstad, to avoid confusion with the marketing company. The red hull livery was adopted from Slite's Ålandspilen service (taken from the colour of the company chairman's wife's lipstick). In 1967, Rederi Ab Ålandsfärjan changed its name to SF Line, and in 1977, Rederi Ab Solstad was merged into its mother company Rederi Ab Sally. Because Viking Line was only a marketing company, each owner company retained their individual fleets and could choose on which routes to set their ships. Each company's ships were easy to distinguish by name: all Sally ships had a 'Viking' prefix to their names, Slite took their names from Roman and Greek mythologies, whilst SF Line's names ended with -ella in honour of the managing director's wife, Ellen Eklund (Figure 16.3).

During the 1970s, Viking expanded and overtook Silja Line as the largest shipping consortium on the Northern Baltic Sea. Between 1970 and 1973, Slite and Sally took delivery of five identical ships built at Meyer Werft Germany, namely *MS Apollo* and *MS Diana* for Slite, and *MS Viking 1*, *MS*

Figure 16.3 Viking Line cruise ferry *MS Viking Cinderella* departing Stockholm, Sweden.

Viking 3 and *MS Viking 4* for Sally. *MS Viking 5*, delivered in 1974, was an enlarged version of the same design. These so-called Papenburg sisters are considered one of the most successful ship designs of all time, with the shipyard building a further three additional sisters of the original design for Transbordadores for service in Mexico: *Coromuel*, *Puerto Vallarta*, and *Azteca*. In 1973, Viking Line started service on the Turku–Mariehamn–Stockholm route, directly competing with Silja Line for the first time. The next year, Sally began Viking Line traffic between Helsinki and Stockholm. For the next decade, this route stayed separate, whereas, on all other routes, the three companies operated together. By the latter half of the 1970s, Sally was clearly the dominant partner in the consortium. In 1980, they took delivery of three new ferries (*MS Viking Saga*, *MS Viking Sally*, and *MS Viking Song*), the largest ships to have sailed under Viking's colours. This further established their dominance over the other partners, although SF Line did take delivery of the new *MS Turella* and *MS Rosella* in 1979–1980 and Slite's *MS Diana II* in 1979. In the early 1980s, Sally started expanding their operations to other operational areas. This would become the company's failure, as those operations proved unprofitable and made Sally unable to invest in new tonnage for the Viking Line service. In 1985, Viking's brand-new *MS Mariella*, at that time the largest ferry in the world, replaced *MS Viking Song* on the Helsinki–Stockholm service, breaking Sally's monopoly on the route. The next year, Slite took delivery of *MS Mariella*'s sister, *MS Olympia*, forcing Sally out of the Helsinki to Stockholm route completely. While SF Line and Slite were planning additional newbuilds, Sally was in an extremely poor position financially, and in 1987, Effoa and Johnson Line, the owners of Silja Line purchased Sally. As a result, SF Line and Slite forced Sally to leave the Viking Line consortium in 1988. Between 1988 and 1990 SF Line took delivery of three new ships (*MS Amorella*, *MS Isabella*, and *MS Cinderella*) while Slite took delivery of two (*MS Athena* and *MS Kalypso*). Unfortunately, Wärtsilä Marine, the shipyard building one of SF Line's newbuilds and both of Slite's, went bankrupt in 1989. SF Line avoided the financial repercussions of Wärtsilä's bankruptcy as their vessel, the *MS Cinderella*, had been continuously paid for as her construction progressed. Hence, it was SF Line who owned the almost completed ship when the shipyard went bankrupt. Slite however had signed a more traditional type of contract; the *MS Kalypso* was to be paid for on delivery. Since the shipyard owned the unfinished ship, this led to an increased cost for the *MS Kalypso*, approximately 200 million SEK more than had been originally envisaged. In the end, despite the financial problems, by 1990, Viking Line had the largest and newest cruise ferry fleet in the world.

In 1989, Slite started planning the *MS Europa*, which was to be the jewel in the company's crown; she was to be the largest and most luxurious cruise ferry in the world. Unfortunately for them, Sweden entered a monetary crisis during the construction of the ship, which led to the devaluation of the Swedish krona (SEK). This in turn meant that the cost for the *MS Europa* increased by 400 million SEK. When the time came to take delivery of the new ship, Slite did not have the funds to pay for her and their main funders (the Swedish bank Nordbanken, who were also the main funders of Silja Line) refused to loan the money needed to purchase the ship. Eventually, the ship ended up in Silja Line's fleet and Slite was forced to declare bankruptcy in 1993. Following the bankruptcy of Rederi AB Slite, SF Line was left as the sole operator under the Viking Line brand. The remaining two Slite ships, *MS Athena* and *MS Kalypso*, were auctioned in August 1993. SF Line made a bid for the *MS Kalypso*, but both ships ended up sold to the newly established Malaysian cruise ship operator Star Cruises. In 1995, SF Line changed their name to Viking Line. Between 1994 and 1996, the company operated a fast ferry service from Helsinki to Tallinn during the summers on chartered catamaran ships. In 1997, they purchased *MS Silja Scandinavia* from Sea-Link Shipping AB and renamed her *MS Gabriella* for the Helsinki–Stockholm service. It was reported that around the same time plans were made to construct a pair of new ships for the Helsinki–Stockholm service so that Viking could better compete with Silja on that route, but these plans were shelved. In 2006, Sea Containers Ltd, who had become the main owner of Silja Line in 1999, put Silja Line and their cargo-carrying subsidiary SeaWind Line for sale, except for the *GTS Finnjet* and *MS Silja Opera*. Both ships were transferred to Sea Container's direct ownership and were eventually sold. In a seminar held in January 2010, Viking Line began negotiations with nine different shipyards about the possibility of constructing a pair of 60,000-tonne ships to replace the *MS Amorella* and *MS Isabella* on the Turku–Stockholm service. The possibility of using liquefied natural gas engines and other emission-reducing technologies was researched, whilst the ships would include various features akin to those found onboard cruise ships such as Royal Caribbean International's *MS Oasis of the Seas*. In October 2010 Viking Line signed a letter of intent with STX Turku for a 57,000 metric-tonne cruise ferry for the Turku–Stockholm route. Two months later, the formal order for the new ship was placed. The new ship christened *Viking Grace*, was laid down on 6 March 2012 and launched on 10 August 2012. The ship entered service in January 2013. Viking Line had an option for a sister ship but announced in May 2012 that they have decided not to build it. Viking Line revealed in November 2016 that a letter of intent had been signed with Chinese shipyard Xiamen Shipbuilding for the construction of a 63,000 metric-tonne cruise ferry that would on completion replace the *MS Amorella* in the Viking Line fleet.

English Channel and Bay of Biscay

P&O

The British shipping company P&O is credited with establishing the first ferry services in the UK in the late 1960s in the North Sea and the English

Channel. In the late 1970s, P&O was affected by a reduction in traditional shipping activities, which saw the sale of several of its businesses and assets. This continued into 1985 with the sale of its cross-channel ferry activities to European Ferries, which at the time consisted of services on the Port of Dover–Boulogne and Southampton–Le Havre routes. In January the following year, P&O purchased a 50.01% interest in European Financial Holdings Ltd, which held 20.8% of shares in European Ferries, followed in 1987 with the purchase of the remaining shares of the European Ferries Group whose ferry services were trading as Townsend Thoresen. Following the *Herald of Free Enterprise* disaster in March 1987, the operations of Townsend Thoresen were renamed P&O European Ferries on 22 October 1987, with operations from Portsmouth, Felixstowe, and Dover. Following a consultation with the Competition Commission beginning on 28 November 1996, P&O European Ferries split into three separate subsidiaries: P&O Portsmouth, P&O North Sea, and the creation of a joint venture between P&O and the Swedish ferry company Stena Line's UK subsidiary Stena Line (UK) Ltd to create P&O Stena Line in Dover. In April 2002, P&O announced its intention to purchase Stena Line's 40% share of the joint venture. The purchase was completed by August, and in October 2002, the Portsmouth and North Sea operations were merged with the Dover operations to create P&O Ferries Ltd, jointly managing all services from its head office, Channel House in Dover. In September 2004, P&O Ferries Ltd conducted a business review that concluded with the announcement of the closure of several of its long-term Portsmouth-based routes, leaving only the Portsmouth – Bilbao route in operation. These closures were blamed on the expansion of low-cost airlines and the increasing usage of the Channel Tunnel as a faster alternative to ferry operations. In 2006, the P&O Group, including P&O Ferries, was sold to Dubai-based DP World. Shortly afterwards, it was taken over by Dubai World. On 15 January 2010, P&O Ferries announced that it would be closing the Portsmouth – Bilbao route by the end of September to coincide with the end of its existing charter for the *Pride of Bilbao*. This meant that the closure of the final route served by P&O Ferries in Portsmouth. In January 2019, P&O Ferries announced that its British fleet would be reflagged from Dover to Limassol, Cyprus, in response to Britain's exit from the EU in 2019. The reason given was 'for operational and accounting reasons'. Cyprus is a member of the EU and a Flag of Convenience; reflagging P&O's vessels would allow P&O to continue to benefit from EU tax arrangements. In response, the Rail, Maritime and Transport (RMT) union condemned the reflagging of P&Os fleet as 'pure opportunism', saying that the company's 'long-term aim has always been to switch the UK fleet to a tax haven register'. On 20 February 2019, DP World announced it had repurchased P&O Ferries from Dubai World in a deal worth £322m (Figure 16.4).

P&O's Portsmouth operations began with their acquisition of Normandy Ferries' Portsmouth – Le Havre route, branding it as P&O Normandy Ferries and offering a twice-daily service, initially competing with both

Figure 16.4 P&O Ferries, English Channel between Dover, England and Calais, France.

Brittany Ferries and Townsend Thoresen. In 1985, P&O sold its ferry operations to European Ferries before returning to the market in 1987 with its takeover of European Ferries that same year, this time employing the P&O European Ferries brand. Following the acquisition in 1987, P&O European Ferries operated routes from Portsmouth to Cherbourg and Le Havre. The Cherbourg route was operated by the *Super Viking class* of vessels, namely the *Pride of Cherbourg (1)* and *Pride of Winchester* until 1994 when they were replaced by the two jumbo-sized 'Super Vikings', *Pride of Cherbourg (2)* and *Pride of Hampshire*. On the Le Havre route, the jumbo *'Super Vikings' Pride of Hampshire* and *Pride of Le Havre* initially serviced the route supported at various times by chartered in freight vessels or transferred 'European' class vessels. However, their low capacity meant that by 1991, larger vessels were required; this was eventually solved by the introduction of two German-built vessels, *Pride of Le Havre* (2) and the in 1994. In 1993, P&O opened a new route from Portsmouth to Bilbao, Spain, using the *Pride of Bilbao* operating a twice-weekly service. This was the longest route that P&O Ferries operated and was often late arriving due to the weather conditions in the Bay of Biscay. The *Pride of Bilbao* was an archetypical cruise ferry. Built for Viking Line, the *Pride of Bilbao* was operated by P&O Ferries between Portsmouth and Bilbao from 1993 to 2010. The Bilbao route remained the only Portsmouth operation, until 15 January 2010, when P&O Ferries announced they would withdraw the service at the end of *Pride of Bilbao's* charter. The vessel completed her final voyage on 28 September 2010 and was returned to Irish Continental Group, from whom she had been chartered since the route's inception. This marked the end of P&O's operations from Portsmouth.

Brittany Ferries

Brittany Ferries is the trading name of the French shipping company, BAI Bretagne Angleterre Irlande SA and was founded in 1973 by Alexis Gourvennec. The company operates a fleet of cruise ferries between the UK, Ireland, and Spain, and between Spain and Ireland and the UK. Collaborating with fellow Breton farmers, Gourvennec lobbied for improvements to Brittany's infrastructure, including better roads, a telephone network, education, and port access. By 1972, he had successfully secured funding and work to develop a deep-water port at Roscoff. Although Gourvennec had no desire to run a ferry service, existing operators showed little appetite for the opportunity. The company began operations on 2 January 1973 between Roscoff in Brittany and Plymouth in the southwest of England, using the freight ferry *Kerisnel*, a former Israeli tank carrier. The company's primary aim at that time was to exploit the opportunities presented by Britain's entry into the European Common Market, the forerunner to the EU, to export directly to markets in the UK. In 1974, *Kerisnel* was replaced by *Penn-Ar-Bed*, which carried both passengers and vehicles. It was at this time the BAI company adopted the name Brittany Ferries. In late 2009, the new Poole – Santander freight-only service was deemed a success and the frequency of service was doubled with two services a week operated by *Cotentin*. In November 2009, *Armorique* was laid up, with major changes to the fleet announced in December 2009. *Barfleur* was withdrawn from service at the end of January 2010 after 18 years of service on the Poole–Cherbourg route. The service was temporarily served by *Armorique*, which came back into service earlier than originally planned. The Poole–Santander service reverted to one sailing a week with *Cotentin* covering freight on the Poole–Cherbourg service in the absence of *Barfleur*. *Condor Vitesse* continued to operate one round sailing a day in the summer months between the two ports. *Cap Finistère* ran the route between Portsmouth and Santander twice a week and operated three round trips a week between Portsmouth and Cherbourg. In September 2010, Brittany Ferries announced plans to serve the Portsmouth–Bilbao route recently abandoned by P&O Ferries. The route started on 27 March 2011 (Figure 16.5).

On 21 September 2012, Brittany Ferries cancelled sailings indefinitely following two days of wildcat strikes caused by crew members who were unhappy with changes in working terms and conditions. Meetings took place between management and unions to negotiate the management proposals. A vote was taken on 30 September by union members to decide if the

Figure 16.5 Brittany Ferries' *Barfleur* leaving Poole, England with Girlfriend in the foreground.

management proposals would be accepted. The crew members accepted the proposal and services resumed on 2 October after 12 days without services. During this period, Brittany Ferries made special arrangements with P&O Ferries and MyFerryLink to accept tickets on the Dover–Calais route. Unused tickets were refunded. Services on the Poole – Cherbourg route were not affected as these were being operated by Condor Ferries. In 2018, Brittany Ferries commenced service between Cork, Ireland, and Santander. This was cancelled and effectively replaced in February 2020 by the Rosslare–Bilbao service which runs twice weekly, with a seasonal service between Rosslare and Roscoff now offered. From late March 2020, due to the ongoing COVID-19 pandemic, Brittany Ferries was forced to cancel all passenger sailings until 15 May 2020 after British government advice was issued against all travel. On 23 July 2020, Brittany Ferries announced the launch of a new Rosslare–Cherbourg service. On 19 August 2020, because of the ongoing COVID crisis, the company confirmed it was reducing services from the end of August and laying up various ships starting with *Armorique* and *Bretagne*. On 20 July 2021, Brittany Ferries announced at a press conference in Paris that it had secured a charter with Stena RORO for two E-Flex class vessels. The new vessels were to replace the *MV Normandie* on the Portsmouth–Caen route and *MV Bretagne* on the Portsmouth – St Malo Route. The charter is expected to run for 10 years with the option to purchase after 4 years.

Stena Line

In 1998, Stena's operations from Dover and Newhaven were merged with P&O European Ferries to form P&O Stena Line, 40% of which was owned by Stena and 60% by P&O. In 2002, P&O acquired all of Stena's shares in the company, thus becoming the sole owner of P&O Stena Line, which soon changed its name to P&O Ferries.

North Sea

P&O

P&O's involvement in the North Sea ferry routes began with a 35% stake in North Sea Ferries owned by its subsidiary, the General Steam Navigation Company. North Sea Ferries began operations on 17 December 1965 sailing on the Hull (England) to Rotterdam (Netherlands) route, a route which critics predicted would not survive. The numbers proved them wrong, however, and in the first year, 54,000 passengers were carried. By 1974, demand for capacity was greater than could be supplied, and two vastly bigger vessels, *Norland* and *Norstar*, which were at that time the largest ferries in the world, were introduced to the route. The two vessels *Norwind* and *Norwave* were transferred to a new route, Hull (England) to Zeebrugge (Belgium) operating a nightly service departing 30 minutes after the Rotterdam service set sail. The two routes remained unchanged, except for the Ministry of Defence chartering *Norland* for service with the British Task Force to the Falkland Islands in 1982. By the mid-1980s, the two routes were becoming increasingly popular, and in 1987, the larger *Norsun* and *Norsea* were introduced on the Hull – Rotterdam route, with the two displaced vessels moving to the Hull – Zeebrugge route and the sale of both *Norwind* and *Norwave*. As part of the success of the routes, dedicated freight routes were introduced from Teesport to both Rotterdam and Zeebrugge on a nightly basis. However, the ships quickly began to struggle to cope with demand, and in 1994, new super freighters were introduced on the Hull routes. A new river berth was therefore constructed to accommodate the freight-only vessels; however, they were still small enough to pass through the lock at King George Dock if required. By 1996, P&O owned a 50% stake in North Sea Ferries, with the other 50% owned by the Dutch shipping company, Royal Nedlloyd Group. In a multimillion-pound deal, P&O purchased Royal Nedlloyd Group's stake, and North Sea Ferries was rebranded to P&O North Sea Ferries. Continuing success saw the purchase of two new super ferries, each weighing 60,000 metric tonnes from Fincantieri, entering service in 2001 as the *Pride of Rotterdam* and the *Pride of Hull* and once again holding the title of the 'world's largest ferry'. *Norsea* and *Norsun* were refitted and returned to P&O North Sea Ferries on the Hull – Zeebrugge route as the *Pride of York* and *Pride of Bruges*. Following P&O's acquisition of P&O Stena Line in 2002, P&O North Sea Ferries was merged and rebranded with P&O's Portsmouth and Dover operations under the current P&O Ferries Ltd name (Figure 16.6).

Figure 16.6 P&O North Sea Ferries *Pride of York* leaving the Humber Estuary, England.

Stena Line

Stena Line is a Swedish shipping line company and one of the largest ferry operators in the world. The company services routes in and around Denmark, Germany, Ireland, Latvia, the Netherlands, Norway, Poland, Sweden, and the UK. Stena Line is a major operating unit of Stena AB, itself a part of the Stena Sphere. Stena Line was founded in 1962 by Sten A. Olsson in Gothenburg, Sweden, which still serves as the company's headquarters, when he acquired Skagenlinjen, a local service operating between Gothenburg and Frederikshavn, Denmark. In 1972, Stena Line was one of the first ferry operators in Europe to introduce a computer-based reservation system for the travel business area. In 1978, the freight business area also started operating a computer-based reservation system. During the 1980s, Stena acquired three other ferry companies starting in 1981 with the Sessan Line, Stena's biggest competitor on lucrative Sweden to Denmark routes. The company was acquired and incorporated into Stena Line. The acquisition included Sessan's two large newbuilds, *Kronprinsessan Victoria* and *Prinsessan Birgitta*, which became the largest vessels operated by Stena at that time. In 1983, Stena acquired the Varberg-Grenå Linjen, and two years later also, the right to that company's former name, Lion Ferry. Lion Ferry continued as a separate marketing company until 1997 when it was fully incorporated into Stena Line. In 1989, Stena acquired yet another ferry company, Stoomvaart Maatschappij Zeeland (SMZ), which at the time traded under the name Crown Line. SMZ's Hook of Holland to Harwich route was fully absorbed into Stena Line, which continues to operate to this day. In 2000, Stena Line purchased yet another Scandinavian ferry operator: Scandlines AB. In November 2006, Stena ordered a pair of 'super ferries' with a gross tonnage of 62,000 dwt from Aker Yards, Germany, which were delivered in 2010, with an option for two more ships of the same design. The new ferries are amongst the largest in the world and are operated on Stena's North Sea route from the Hook of Holland to Harwich. The existing ships on the North Sea route were transferred to the Kiel (Germany) – Gothenborg (Sweden) route, whereas the ships from Kiel would transfer to the Gdynia (Poland) to Karlskrona (Sweden) route. These new ferries were launched in 2010, with *Stena Hollandica* entering service on 16 May 2010, and *Stena Britannica* entering service on 9 October 2010 (Figure 16.7).

DFDS Seaways

DFDS Seaways is a Danish shipping company that operates passenger and freight services across northern Europe. Following the acquisition of Norfolkline in 2010, DFDS restructured its other shipping divisions (DFDS Tor Line and DFDS Lisco) into the previously passenger-only operation of DFDS Seaways. DFDS Seaways renewed its fleet in 2006, purchasing

Figure 16.7 Stena Line ferry *Stena Hollandica*, departing Harwich, England.

MS King of Scandinavia and *MS Princess of Norway* to replace the last ships still in service that dated from the 1970s. The company has acquired a reputation for purchasing used ships, as well as for taking over the build contracts or taking delivery of newbuilds originally ordered by other companies. The last time DFDS Seaways ordered a newbuild of its own was in 1978. DFDS Seaways stopped serving Sweden in 2006 when *MS Princess of Scandinavia* was taken out of service and the Copenhagen to Oslo service stopped calling at Helsingborg (Sweden). In May 2008, DFDS made its plan public to close the loss-making England to Norway service on 1 September 2008. *MS Queen of Scandinavia*, the ship that was used for that service, has since been chartered to St Peter Line. In July 2010, DFDS acquired Norfolkline from Mærsk, after which the Norfolkline routes and vessels were fully integrated into DFDS Seaways. Following the acquisition of Norfolkline, DFDS Seaways now controlled the Dover to Dunkirk (France) route. The company launched the new Dover to Calais (France) route in February 2012. In 2018, DFDS ordered three ROPAX newbuild vessels – one was an *E-Flex class* on charter from Stena RORO, which entered service as the *Côte d'Opale* in August 2021, and two ROPAXs for Baltic Sea operations. The first of the Baltic twins, named *Aura Seaways*, was launched in late 2020 and had an inclination test in 2021. The sea trials took place in the middle of 2021 (Figure 16.8).

Figure 16.8 DFDS ferry *Princess Seaways*.

Irish Sea

P&O

In 1971, P&O purchased the remains of Coast Lines, which had been operating in the Irish Sea since 1913. In December 1974, P&O founded Pandoro Ltd to provide transport operations to Ireland. Several different routes were started and ceased operating as the Irish 'Troubles' affected the car and passenger market to Northern Ireland. In 1993, Pandoro added a service operating between Rosslare, Ireland, and Cherbourg, France, to its Ardrossan – Larne, Liverpool – Dublin and Fleetwood – Dublin routes. P&O Irish Sea was formed in 1998, following the merger of the Cairnryan-based service of P&O European Ferries (Felixstowe) Ltd and Pandoro Ltd (who operated routes between England, Scotland and France to Ireland). The following year in 1999, the new P&O Irish Sea announced its intentions to purchase a purpose-built ROPAX vessel from Mitsubishi Heavy Industries of Japan for the Liverpool to Dublin route. This would see the transfer of the *European Leader* (ex-*Buffalo*) back to the Fleetwood route. In 2004, P&O closed its Fleetwood to Larne services with the sale of all interests. In addition to the service rights, *European Leader, European Pioneer*, and *European Seafarer* were sold to the Swedish Stena Line group. At the same time, P&O announced the closure of the Mostyn to Dublin service due to low passenger numbers. This led to the sale of *European Ambassador* and *European Envoy* for further service in Europe. In 2010, P&O Irish Sea was absorbed back into the parent company. In January 2016, it was announced that the seasonal Troon to Larne service would cease with immediate effect due to the unprofitability of the route. The service ended for the 2015 season in September of that year (Figure 16.9).

Stena Line

In 1990, Stena Line doubled in size with the acquisition of Sealink British Ferries from Sea Containers. This first became Sealink Stena Line, then Stena Sealink Line, and finally Stena Line (UK), which now operates all Stena's ferry services between Great Britain and Ireland (Figure 16.10).

Scotland

NorthLink Ferries

NorthLink Ferries (also referred to as Serco NorthLink Ferries) is an operator of passenger and vehicle ferries, as well as ferry services, between mainland Scotland and the Northern Isles of Orkney and Shetland. Since July 2012, it has been operated by the international services company, Serco. The subsidised Northern Isles ferry services, previously run by P&O Scottish Ferries, were put out to tender in 1999. A joint venture between Caledonian MacBrayne and The Royal Bank of Scotland, named NorthLink Orkney and

Figure 16.9 P&O Irish *Sea Express.*

Figure 16.10 Stena Super fast X, arriving in Dublin, Ireland.

Shetland Ferries, won the contract and began operation in October 2002. A variety of factors, including competition from rival operator Pentland Ferries, the Norse Island Ferries group created by local hauliers concerned about NorthLink's proposed freight pricing, and higher-than-expected operation costs, contributed to financial difficulties within the company. In response, the Scottish Executive Transport Group (now Transport Scotland) made additional subsidy payments of £0.6 million and agreed to restructure subsidy payment timing. In mid-2003, the company indicated that it would be unlikely to complete its contract due to the ongoing financial difficulties. NorthLink defaulted on its lease payments for the vessels in July and August 2003, and in April 2004, the then Scottish Executive announced that the service would be re-tendered due to NorthLink's inability to fulfil the terms of its contract. The company continued to operate under interim arrangements until April 2006, whilst a new contract was secured. On 19 July 2005, the Scottish Executive announced that three companies – V Ships, Irish Continental Ferries, and Caledonian MacBrayne, had bid to provide ferry services to the Northern Isles. Irish Continental, however, withdrew its bid in October 2005, leaving two potential operators on the closing date of 1 December 2005. Both the remaining bids complied with the contract requirements, but Caledonian MacBrayne's lower bid meant that it was awarded the contract. Caledonian MacBrayne formed a company named NorthLink Ferries Limited, which adopted the branding and vessels of its predecessor and began operating the Northern Isles ferry services on 6 July 2006. The Northern Isles ferry service was re-tendered in 2011–2012 as NorthLink Ferries Limited's contract ended. Initially, the contract's two services (Aberdeen to Lerwick and Aberdeen/Lerwick to Scrabster-Stromness) were to be de-bundled. Eligible bids for the services were received from Pentland Ferries (which expressed interest in the Scrabster-Stromness service only), Sea-Cargo A/S (which expressed interest in the Aberdeen-Lerwick service only), P&O Ferries, Shetland Line (1984) Limited (part of local haulage and freight company Streamline Shipping Group), Serco, and the incumbent NorthLink Ferries Limited. The Scottish Government subsequently re-bundled the routes when insufficient interest was shown in the separate routes (Figure 16.11).

On 4 May 2012, Transport Scotland announced that Serco was the preferred bidder. This decision was legally challenged in the Court of Session by rival bidder Shetland Line (1984) Limited on the basis that the Scottish Government had allegedly not considered that they had scored higher than Serco for their proposed service, suspending the securement of the contract. On 29 May 2012, however, the court overturned the suspension and Serco was confirmed as the new operator, ending Caledonian MacBrayne's 10-year involvement with Northern Isles ferry services. Each contract lasts for a period of six years and is worth £243 million. Serco, using the vessels and branding of its predecessor, began operation of Northern Isles ferry services at 15:00 on 5 July 2012. It stated that it planned to make no changes to fares or timetables for the remainder of 2012 and that it planned to

Figure 16.11 North Link Ferries *Hrossey* on route to the Shetland Isles, Scotland.

'overhaul catering, seating and onboard entertainment' in the future. In Spring 2013, NorthLink rebranded and launched new on-board services such as 'sleep pod' reclining seats and a premium lounge. The contract was due to end in 2018, but Serco received an 18-month extension. Arguments have been put forward by the RMT union to bring the service into public ownership. In February 2020, Paul Wheelhouse, the Scottish Minister for Energy, Connectivity and the Islands, announced that NorthLink's contract would be re-awarded by the end of March of that year.

Skagerrak

Color Line

Color Line AS is the largest cruise ferry line operating on routes to and from Norway. The company is also one of the leading ferry operators in Europe. Color Line has roots in the ferry business that go back more than one hundred years, although the present company was only established in 1990 when two Norwegian shipping companies, Jahre Line and Norway Line merged. Jahre Line had operated ferries between Oslo (Norway) and Kiel (Germany) since 1961, whilst Norway Line had operated ferries from Norway to the UK and the Netherlands since 1986. During 1990, Color Line also took over the Fred. Olsen Lines cruise ferry operations, thereby expanding the traffic area of the new company to cover the Norway to Denmark routes. During the first half of the 1990s, Color Line expanded its tonnage by lengthening its existing ships or by the acquisition of larger second-hand ships. The company began operating fast ferries between Norway and Denmark during the summer of 1996. Initially, the operations were in collaboration with SeaContainers but were run without them from 1997 onwards.

Figure 16.12 Superspeed 1 arriving at Hirtshals Harbour, Denmark.

In October of the same year, Color Line took over the operations of Larvik Line, its competitor on the Norway to Denmark route. In September 1998, Color Line acquired both the Color Hotel Skagen and Scandi Line, which operated two ferries on the short routes connecting Norway and Sweden. Towards the end of 1998, the Norway – England operations were sold to Fjord Line. For the 1999 summer season, the (former) Scandi Line ships received new Color Scandi Line liveries. They were fully incorporated into the Color Line fleet in 2001 (Figure 16.1).

During the 2000s, Color Line began investing heavily in new tonnage, with *MS Color Fantasy*, *MS Color Magic*, *MS Superspeed 1*, and *MS Superspeed 2* supplanting much of the older tonnage between 2004 and 2008. In April 2008, the company announced the closure of the Oslo (Norway) to Hirtshals (Denmark) service from 6 May 2008 onwards. In January 2017, Color Line announced that it had signed a letter of intent with the Ulstein Verft shipyard to build a new ferry with a hybrid drivetrain for the Sandefjord (Norway) to Strömstad (Norway) route, with delivery expected in 2019. The keel was laid in April 2018 in Poland and the ship was launched in November 2018.

Mediterranean

Grimaldi Lines

The Italian company Grimaldi Group employs around 30 ROPAX vessels for the mixed transport of goods and passengers in the Mediterranean Sea, the Baltic Sea, and the North Sea. With a fleet consisting mostly of modern and comfortable ferries, cruise ferries, and a luxury catamaran, the Group transports rolling freight, cars, and passengers between the main European ports under the Grimaldi Lines, Minoan Lines and Finnlines brands. Fleet improvement initiatives have involved the two flagships of the Group's ROPAX fleet. The two state-of-the-art cruise ferries *Cruise Roma* and *Cruise Barcelona* – both operating since 2008 – underwent major lengthening and refurbishment works at the beginning of 2019, which further increased their energy efficiency and made them the first ships in the Mediterranean with Zero Emission in Port technology. In fact, thanks to the installation of mega lithium-ion batteries that are recharged during navigation, the vessel can cut emissions to zero during port stays. Each of the sister vessels can currently carry up to 3,500 passengers, 3,700 linear metres of rolling cargo, and 271 cars. In December 2019, Finnlines finalised a new order for the construction of two ROPAX units, which are expected to be delivered by 2023. These new ships, which will inaugurate the innovative Superstar class, will be larger and more technologically advanced than the existing units belonging to the Star class. This investment is aimed at increasing energy efficiency and further reducing the emissions generated by transport activities in the Baltic Sea, while at the same time raising the quality of the services offered to the company's passengers (Figure 16.13).

Moby Lines

Moby Lines (Moby Lines SpA) is an Italian shipping company that operates ferries and cruise ferries between the Italian or French mainland and the islands of Elba, Sardinia, and Corsica. The company was founded in 1959 under the name Navigazione Arcipelago Maddalenino (often abbreviated to NAVARMA). In 2006, Moby Lines purchased Lloyd Sardegna, which was known for using Warner Bros. Looney Tunes characters as the external livery of its ships. NAVARMA was founded in 1959 by Achille Onorato and

Figure 16.13 Grimaldi Lines *Finclipper* in Brindisi, Italy.

started traffic from Sardinia to the islands on the coast of Sardinia with the small ferry *MS Maria Maddalena*, which was purchased from Denmark. In February 1966, NAVARMA purchased a second ferry, *MS Bonifacio*, and started service between Sardinia and Corsica. The company slowly expanded, purchasing another ferry in 1967 and taking delivery of two newbuilds in 1974 and 1981. With the larger fleet, new routes to the Italian mainland were also introduced. In 1982 the company acquired the *MS Free Enterprise II* from Townsend Thoresen, and renamed her the *MS Moby Blu*, painting her in the 'blue whale' livery that later came to characterise Moby Lines. The *Moby Blu* was over twice the size of NAVARMA's previously largest ship. By 1988 four additional larger ferries (all with Moby-prefixed names) had joined the NAVARMA fleet and additional routes to the Italian mainland were opened. In 1991 one of the ferries of the fleet, the *Moby Prince*, participated in the worst disaster to befall the Italian merchant navy since World War II, resulting in 140 deaths. By comparison, the *Costa Concordia* disaster in 2012 caused 32 fatalities. During the early 1990s, NAVARMA acquired further used ferries, which replaced the Moby ferries acquired in the 1980s. During the same time, 'Moby Lines' was adopted as the official company name. From 1996 onwards, the company fleet had grown radically with the addition of new, larger and faster tonnage, including the newly built fast cruise ferries *Moby Wonder*, *Moby Freedom*, and *Moby Aki*. In 2003, Moby Lines entered an agreement with Warner Bros. to paint their vessels in liveries featuring Looney Tunes characters. However, only the larger ships were given such liveries, with the company's smaller ships either having similar graphics not featuring the Looney Tunes characters, or simply featuring the Moby Lines' whale logo. In 2020, in a departure from Moby Lines' signature business model of acquiring vintage tonnage for its routes, it was announced that steel cutting had started for Moby Lines' two newbuild vessels on order from the Guangzhou Shipyard. These newbuilds would be 238 m (784 ft) long and 69,500 metric-tonnes dwt, and specifically designed for the 7-to-9-hour Livorno-Olbia ferry crossing. These newbuild twins are earmarked to replace *Moby Aki* and *Moby Wonder* in 2022.

In this chapter, we have discussed the role and function of the cruise ferry, a sort of hybrid type vessel, which is used to transport passengers and vehicles over longer distances than might be expected of the more typical ferry. Because cruise ferries combine many of the luxuries and hospitality traits of cruise ships, but also move cargo from port to port, the cruise ferry occupies a rather odd position in the categories of merchant vessels.

NOTE

1 Snus is a tobacco product, originating from a variant of dry snuff in early 18th-century Sweden. The sale of snus is illegal in all the EU countries except for Sweden. It is the most common type of tobacco product in Norway and is also available in Switzerland.

Chapter 17

Ferries

A ferry is a vessel used to carry passengers, and sometimes vehicles and cargo, across a body of water. A passenger ferry with many stops, such as in Venice, Italy, is sometimes called a water bus or water taxi. Ferries form a part of the public transport systems of many waterside cities and islands, allowing direct transit between points at a capital cost much lower than bridges or tunnels. Ship connections of much larger distances (such as over long distances in water bodies like the Mediterranean Sea) may also be called ferry services, especially if they carry vehicles. The concept of the ferry dates from antiquity, as far back as ancient Greece (12th to 9th centuries BC). The profession of the ferryman is embodied in Greek mythology in Charon, the boatman, who transported souls across the River Styx to the Underworld. Speculation that a pair of oxen propelled a ship having a water wheel can be found in the fourth century in the Roman literature 'Anonymus De Rebus Bellicis'. Although impractical, there is no reason it could not work and such a ferry, the *Experiment*, modified by using horses, was used on Lake Champlain in 19th-century America. In 1871, the world's first ferry ship was designed in Istanbul. The iron steamship, *Suhulet* (meaning 'ease' or 'convenience') was designed by the general manager of Şirket-i Hayriye (Bosporus Steam Navigation Company), Giritli Hüseyin Haki Bey and built in Britain. The vessel weighed 157 metric tonnes, had a length overall of 45.7 m (155 ft), a beam of 8.5 m (27 ft), and a draught of 3 m (9 ft). She could sail at six knots (6.9 mph, 11.11 km/h) with the side wheel turned by its 450 horsepower, single-cylinder, two-cycle steam engine. Launched in 1872, Suhulet's unique features consisted of a symmetrical entry and exit for horse carriages, along with a dual system of hatchways. The ferry operated on the Üsküdar-Kabataş route, which is still serviced by modern ferries today (Figure 17.1).

The world's largest ferries are typically those operated in Europe, with different vessels holding the record depending on whether length, gross tonnage, or car vehicle capacity is the metric. In the USA, the first steam-ship ferry started operation between New York City, New York, and Hoboken, New Jersey. This service began on 11 October 1811, using inventor John Stevens' ship the *Juliana*. The oldest American ferry service still in operation

DOI: 10.1201/9781003342366-21

Figure 17.1 Excelsior at Séte, France.

today is the *Elwell Ferry*, a cable ferry in North Carolina, which began in 1905, and travels 100 m (110 yds), shore to shore, with a travel time of 5 minutes. A contender for the oldest ferry service in continuous operation is the Mersey Ferry from Liverpool to Birkenhead, England. In 1150, the Benedictine Priory at Birkenhead was established. The monks used to charge a small fare to row passengers across the estuary. In 1330, Edward III granted a charter to the Priory and its successors forever: 'the right of ferry there… for men, horses, and goods, with leave to charge reasonable tolls'. However, there may have been a short break following the Dissolution of the monasteries after 1536. Another claimant as the oldest ferry service in continuous operation is the Rocky Hill – Glastonbury Ferry, running between the towns of Rocky Hill and Glastonbury, Connecticut. Established in 1655, the ferry has run continuously since, only ceasing operation every winter when the river freezes over. A long-running saltwater ferry service is the Halifax/Dartmouth ferry, running between the cities of Halifax and Dartmouth, Nova Scotia, which has run year-round since 1752, and is currently managed by the region's transit authority, Metro Transit. That said, the Mersey Ferry predates it as the oldest saltwater ferry.

The busiest seaway in the world, the English Channel, connects Great Britain and mainland Europe, with ships sailing from the British ports of Dover, Newhaven, Poole, Portsmouth, and Plymouth to French ports, such as Calais, Dunkirk, Dieppe, Roscoff, Cherbourg-Octeville, Caen, St Malo, and Le Havre. The busiest ferry route to France is the Dover to Calais crossing with approximately 9,168,000 passengers using the service in 2018. Ferries from Great Britain also sail to Belgium, the Netherlands, Norway, Spain, and Ireland. Some ferries carry tourist traffic, but most also carry freight, and some are exclusively for the use of freight lorries. In Britain, car-carrying ferries are sometimes referred to as RORO for the ease by which vehicles can board and leave. The busiest single ferry route in terms of the number of departures is across the northern part of Øresund, between Helsingborg, Scania, Sweden, and Elsinore, Denmark. Before the Øresund bridge was opened in July 2000, car and 'car and train' ferries departed up to seven times every hour. This has since been reduced, but a car ferry still departs from each harbour every 15 minutes during the daytime. The route is around 2.2 nautical miles (2.5 mi; 4.1 km) and the crossing takes 22 minutes. Today, all ferries on this route are constructed so that they do not need to turn around in the harbours. This also means that the ferries lack stems and sterns since the vessels sail in both directions. Starboard and port-side are dynamic, depending on the direction the ferry sails. Despite the short crossing, the ferries are equipped with restaurants (on three out of four ferries), cafeterias, and kiosks. Passengers without cars often make a double or triple return journey in the restaurants; for this, a single journey ticket is sufficient. Passenger and bicycle passenger tickets are inexpensive compared with longer routes.

Large cruise ferries sail in the Baltic Sea between Finland, Åland, Sweden, Estonia, Latvia, and Saint Petersburg, Russia, and from Italy to Sardinia, Corsica, Spain, and Greece. In many ways, these ferries are like cruise ships, but they can also carry hundreds of cars on car decks. Besides providing passenger and car transport across the sea, Baltic Sea cruise ferries are a popular tourist destination unto themselves, with multiple restaurants, nightclubs, bars, retail outlets, and entertainment on board. Also, many smaller ferries operate on domestic routes in Finland, Sweden, and Estonia. The southwest and southern parts of the Baltic Sea have several routes for heavy traffic and cars. The ferry routes of Trelleborg-Rostock, Trelleborg-Travemünde, Trelleborg-Świnoujście, Gedser-Rostock, Gdynia-Karlskrona, and Ystad-Świnoujście are all typical transport ferries. On the longer part of these routes, simple cabins are available. The Rødby-Puttgarden route also transports day passenger trains between Copenhagen and Hamburg, and on the Trelleborg-Sassnitz route, it also has capacities for the daily night trains between Berlin and Malmö. In Istanbul, ferries connect the European and Asian shores of the Bosphorus, as well as the Princes Islands and nearby coastal towns. In 2014, İDO transported 47 million passengers, the largest ferry system in the world. The world's shortest ferry line is the Ferry Lina In Töreboda, Sweden. It takes around 20–25 seconds and is hand powered.

Due to the number of large freshwater lakes and length of shoreline in Canada, various provinces and territories have ferry services. BC Ferries operates the third largest ferry service in the world, which carries travellers between Vancouver Island and the British Columbia mainland on the country's west coast. This ferry service operates on other islands including the Gulf Islands and Haida Gwaii. In 2015, BC Ferries carried more than eight million vehicles and 20 million passengers. In Vancouver, SeaBus operates as Canada's east coast inter- and intra-provincial ferry and coastal service, providing a large network operated by the federal government under CN Marine and later Marine Atlantic. Private and publicly owned ferry operations in eastern Canada include Marine Atlantic, serving the island of Newfoundland, as well as Bay, NFL, CTMA, Coastal Transport, and STQ. Canadian waters in the Great Lakes once hosted numerous ferry services, but these have been reduced to those offered by Owen Sound Transportation and several smaller operations. There are also several commuter passenger ferry services operated in major cities, such as Metro Transit in Halifax, and Toronto Island ferries in Toronto. There is also the Société des traversiers du Québec. The *MV Spokane*, sailing from Edmonds to Kingston, is one of the 10 routes served by Washington State Ferries. Washington State Ferries operates the most extensive ferry system in the continental US and the second largest in the world by vehicles carried, with 10 routes on Puget Sound and the Strait of Juan de Fuca serving terminals in Washington and Vancouver Island. In 2016, Washington State Ferries carried 10.5 million vehicles and 24.2 million riders in total. The Alaska Marine Highway System provides service between Bellingham, Washington, and various towns and villages throughout Southeast and Southwest Alaska, including crossings of the Gulf of Alaska. AMHS provides affordable access to many small communities which otherwise would have no road connection or airport access.

The Staten Island Ferry in New York City, sailing between the boroughs of Manhattan and Staten Island, is the nation's single busiest ferry route by passenger volume. Unlike riders on many other ferry services, Staten Island Ferry passengers do not pay any fare to ride it. New York City also has a network of smaller ferries, or water taxis, that shuttle commuters along the Hudson River from locations in New Jersey and Northern Manhattan down to the midtown, downtown and Wall Street business centres. Several ferry companies also offer services linking midtown and lower Manhattan with locations in the boroughs of Queens and Brooklyn, crossing the city's East River. New York City Mayor Bill de Blasio announced in February 2015 that New York City would begin an expanded Citywide Ferry Service some time in 2017 linking the isolated communities of Manhattan's Lower East Side, Soundview in The Bronx, Astoria, and the Rockaways in Queens and the Brooklyn neighbourhoods of Bay Ridge, Sunset Park, and Red Hook. These new services would link into existing ferry landings in Lower Manhattan and Midtown Manhattan. This service was launched as NYC Ferry in May 2017. The New Orleans area also has many ferries in operation that carry both vehicles and pedestrians. Most notable is the Algiers Ferry. This service has been in continuous operation since 1827 and is one of the oldest operating ferries in North America. In New England, vehicle-carrying ferry services between mainland Cape Cod and the islands of Martha's Vineyard and Nantucket are operated by The Woods Hole, Martha's Vineyard, and Nantucket Steamship Authority, which sails year-round between Woods Hole and Vineyard Haven as well as Hyannis and Nantucket. Seasonal service is also operated from Woods Hole to Oak Bluffs from Memorial Day to Labour Day. As there are no bridges or tunnels connecting the islands to the mainland, The Steamship Authority ferries in addition to being the only method for transporting private cars to or from the islands, also serves as the only link by which heavy freight and supplies such as food and fuel can be trucked to the islands. Additionally, Hy-Line Cruises operates high speed catamaran service from Hyannis to both islands, as well as traditional ferries, and several smaller operations run seasonal passenger-only services primarily geared towards tourist day-trippers from other mainland ports, including New Bedford, (New Bedford Fast Ferry) Falmouth, (Island Queen ferry and Falmouth Ferry), and Harwich (Freedom Cruise Line). Ferries also bring riders and vehicles across Long Island Sound to such Connecticut cities as Bridgeport and New London, and to Block Island in Rhode Island from points on Long Island.

Transbay commuting in the San Francisco Bay Area was primarily ferry based until the use of personal vehicles expanded in the 1940s and most bridges in the area were built to supplant ferry services. By the 1970s, ferries were primarily used by tourists with Golden Gate Ferry, an organisation under the ownership as the same governing body as the Golden Gate Bridge, left as the sole commute operator. The 1989 Loma Prieta earthquake prompted restoration of service to the East Bay. The modern ferry network is primarily under the authority of San Francisco Bay Ferry, connecting with cities as far as Vallejo. Tourist excursions are also offered by Blue & Gold Fleet and Red & White Fleet. A ferry serves Angel Island (which also accepts private craft). Alcatraz is served exclusively by ferry service administered by the National Park Service. Until the completion of the Mackinac Bridge in the 1950s, ferries were used for vehicle transportation between the Lower and the Upper Peninsulas of Michigan, across the Straits of Mackinac in the US. Ferry service for bicycles and passengers continues across the straits for transport to Mackinac Island, where motorised vehicles are completely prohibited. This crossing is made possible by three ferry lines, Arnold Transit Company, Shepler's Ferry, and Star Line Ferry. A ferry service runs between Milwaukee, Wisconsin, and Muskegon, Michigan, operated by the Lake Express. Another ferry *SS Badger* operates between Manitowoc, Wisconsin, and Ludington, Michigan, with both crossing Lake Michigan. Mexico has ferry services managed by Baja Ferries that connect La Paz located on the Baja California Peninsula with Mazatlán and Topolobampo. Passenger ferries also run from Playa del Carmen to the island of Cozumel.

In Australia, two Spirit of Tasmania ferries carry passengers and vehicles 280 mi (450 km) across the Bass Strait, the body of water that separates Tasmania from the Australian mainland, often under turbulent sea conditions. These not only run overnight but also include day crossings in peak time. Both ferries are based in the northern Tasmanian port city of Devonport and sail to Melbourne. The double-ended Freshwater-class ferry cuts an iconic shape, as it makes its way up and down Sydney Harbour New South Wales, Australia, between Manly and Circular Quay. In New Zealand, ferries connect Wellington in the North Island with Picton in the South Island, linking New Zealand's two main islands. The route is 57 mi (92 km) and is run by two companies: the government owned Interislander, and the independent company Bluebridge. The passage takes about 3 and a half hours.

FERRY TYPES

Ferry designs depend on the length of the route, the passenger or vehicle capacity required, speed requirements, and the water condition the vessel must deal with.

Double-ended

Double-ended ferries have interchangeable bows and sterns, allowing them to shuttle back and forth between two terminals without having to turn around. Well-known double-ended ferry systems include the BC Ferries, the Staten Island Ferry, Washington State Ferries, Star Ferry, several ferries on the North Carolina Ferry System, and the Lake Champlain Transportation Company. Most Norwegian fjord and coastal ferries are double-ended vessels. All ferries from southern Prince Edward Island to the mainland of Canada were double ended. This service was discontinued upon completion of the Confederation Bridge. Some ferries in Sydney, Australia, and British Columbia are also double ended. In 2008, BC Ferries launched the first of the Coastal-class ferries, which at the time were the world's largest double enders. These were surpassed as the world's largest double-enders when P&O Ferries launched their first double-ender, called the *P&O Pioneer*, which is due to enter service in September 2022 (Figure 17.2).

Figure 17.2 Red Funnel Line's *Red Eagle*, Southampton, England.

Hydrofoil

Hydrofoils have the advantage of higher cruising speeds, succeeding hovercraft on some English Channel routes where the ferries now compete against the Eurotunnel and Eurostar trains that use the Channel Tunnel. Passenger-only hydrofoils also proved a practical, fast, and economical solution in the Canary Islands, but were recently replaced by faster catamaran 'high speed' ferries that can carry cars. Their replacement by the larger craft is seen by critics as a retrograde step given that the new vessels use much more fuel and foster the inappropriate use of cars in islands already suffering from the impact of mass tourism (Figure 17.3).

Hovercraft

Hovercraft was developed in the 1960s and 1970s to carry cars. The largest was the massive *SR.N4*, which carried cars in its centre section with ramps at the bow and stern between England and France. The hovercraft was superseded by catamarans, which are as fast and are less affected by sea and weather conditions. Only one service now remains, a foot passenger service between Portsmouth and the Isle of Wight run by Hovertravel (Figure 17.4).

Catamaran

Since 1990 high speed catamarans have revolutionised ferry services, replacing hovercraft, hydrofoils, and conventional monohull ferries. In the 1990s, there were a variety of builders, but the industry has consolidated to two builders of large vehicular ferries between 60 and 120 m (196–393 ft).

Figure 17.3 Sydney Hydrofoil ferry *Palm Beach* en route from Manly to Circular Quay, Australia.

Figure 17.4 Hoverspeed, leaving Calais, France.

Incat of Hobart, Tasmania, favours a Wave-piercing hull to deliver a smooth ride, while Austal of Perth, Western Australia, builds ships based on SWATH designs. Both these companies also compete in the smaller river ferry industry with several other ship builders. Stena Line once operated the largest catamarans in the world, the Stena *HSS class*, between the UK and Ireland. These waterjet-powered vessels, displaced 19,638 metric tonnes, accommodating 375 passenger cars and 1500 passengers. Other examples of these super-size catamarans are operated by the Condor Ferries fleet with the *Condor Voyager* and *Rapide* (Figure 17.5).

RORO ferries

RORO ferries are large conventional ferries named for the ease by which vehicles can board and leave (Figure 17.6).

Cruiseferry/ROPAX

A cruiseferry is a ship that combines the features of a cruise ship with a roll-on/roll-off ferry. They are also known as ROPAX for their combined roll-on/roll-off and passenger design. Fast ROPAX ferries are conventional ferries with a large garage intake and a large passenger capacity, with conventional diesel propulsion and propellers that sail over twenty-five knots (29 mph; 46 km/h). The pioneer of this class of ferries was the Attica Group, which introduced the *Superfast I* between Greece and Italy in 1995.

Figure 17.5 Condor Rapide, approaching the Port of Guernsey.

Figure 17.6 Jutlandia Seaways, Cuxhaven, Germany.

Turntable ferry

The turntable ferry *MV Glenachulish* (built in 1969), which operates between Glenelg on the Scottish mainland, and Kylerhea, on the Isle of Skye, is the last manually operated turntable ferry in the world. This type of ferry allows vehicles to load from the 'side'. The vehicle platform can be turned.

When loading, the platform is turned sideways to allow sideways loading of vehicles. Then, the platform is turned back, in line with the vessel, and the journey across water is made.

Pontoon and cable ferries

Pontoon ferries carry vehicles across rivers and lakes and are widely used in less-developed countries with large rivers where the cost of bridge construction is prohibitive. One or more vehicles are transported on a pontoon with ramps at either end for vehicles to drive on and off. Cable ferries are usually pontoon ferries, but pontoon ferries on larger rivers are motorised and able to be steered independently like a boat. Noticeably, short distances may be crossed by a cable or chain ferry, which is usually a pontoon ferry, where the ferry is propelled along and steered by cables connected to each shore. Sometimes, the cable ferry is human-powered by someone on the boat. Reaction ferries are cable ferries that use the perpendicular force of the current as a source of power. Examples of a current propelled ferry are the four Rhine ferries in Basel, Switzerland. Cable ferries may be used in fast-flowing rivers across short distances. With an ocean crossing of approximately 1900 m, the cable ferry between Vancouver Island and Denman Island in British Columbia is the longest in the world. Free ferries operate in some parts of the world, such as at Woolwich in London, England (across the River Thames); in Amsterdam, Netherlands (across the IJ waterway); along the Murray River in South Australia, and across many lakes in British Columbia. Many cable ferries operate on lakes and rivers in Canada, among them a cable ferry that charges a toll operates on the Rivière des Prairies between Laval-sur-le-Lac and Île Bizard in Quebec, Canada. In Finland, there were 40 cable ferries in 2009, on lakes, rivers, and between Finland's many islands.

Train ferry

A train ferry is a ship designed to carry railway vehicles. Typically, one level of the ship is fitted with railway tracks, and the vessel has a door at either or both the front and rear to give access to the quayside.

Foot ferry

Foot ferries are small craft used to ferry foot passengers, and often also cyclists, over rivers. These are either self-propelled craft or cable ferries. Such ferries are, for example, to be found on the lower River Scheldt in Belgium and in particular the Netherlands. Regular foot ferry service also exists in the capital of the Czech Republic, Prague, and across the Yarra River in Melbourne, Australia at Newport. Restored, expanded ferry service in the Port of New York and New Jersey uses boats for pedestrians only (Figure 17.7).

Figure 17.7 Gosport Ferry *Spirit of Portsmouth*, seen departing Portsmouth, England, and heading towards Gosport, England.

DOCKING

Ferries often dock at specialised facilities designed to position the ship for loading and unloading, called a ferry slip. If the ferry transports road vehicles or railway carriages, there will usually be an adjustable ramp called an apron that is part of the slip. In other cases, the apron ramp will be a part of the ferry itself, acting as a wave guard when elevated and lowered to meet a fixed ramp at the terminus – a road segment that extends partially underwater or meet the ferry slip.

SUSTAINABILITY

The contributions of ferry travel to climate change have received less scrutiny than land and air transport and vary according to factors like speed and the number of passengers carried. Average carbon dioxide (CO_2) emissions by ferries per passenger kilometre seem to be 0.12 kg (4.2 oz). However, 18-knot (21 mph; 33 km/h) ferries between Finland and Sweden produce 0.221 kg (7.8 oz) of CO_2, with total emissions equalling a CO_2 equivalent of 0.223 kg (7.9 oz), while 24–27 knot (28–31 mph; 44–50 km/h) ferries between Finland and Estonia produce 0.396 kg (14.0 oz) of CO_2 with total emissions equalling a CO_2 equivalent of 0.4 kg (14 oz). With the price of oil at elevated levels, and with increasing pressure from consumers for measures to tackle global warming, several innovations for energy and the environment were put forward at the 2018 Interferry Conference in Stockholm. According to the company Solar Sailor, hybrid marine power and solar wing technology are suitable for use with ferries, private yachts, and even tankers. Alternative fuels are becoming more widespread on ferries. The fastest passenger ferry in the world *Buquebus*, runs on LNG, while since 2015 Sweden's Stena has operated its 1,500 passenger ferries on methanol. Both fuels reduce emissions considerably and replace costly diesel fuel. Megawatt-class battery electric ferries operate in Scandinavia, with several more scheduled for operation. As of 2017, the world's biggest purely electric ferry was the *MF Tycho Brahe*, which operates on the Helsingør–Helsingborg ferry route across the Øresund between Denmark and Sweden. The ferry weighs 8414 metric-tonnes and an electric storage capacity of more than 4 MWh. Since 2015, Norwegian ferry company Norled has operated e-ferry *Ampere* on the Lavik-Opedal connection on the E39 north of Bergen. Further north on the Norwegian west coast, the connection between Anda and Lote was the world's first route served only by e-ferries. The first of two ships, *MF Gloppefjord*, was put into service in January 2018, followed by *MF Eidsfjord*. The owner, Fjord1, commissioned a further seven battery-powered ferries to be in operation from 2020. A total of 60 battery-powered car ferries have been operational in Norway since 2021. Since 15 August 2019, Ærø Municipality has operated e-ferry *Ellen* between the southern Danish ports of Fynshav and Søby, on the island of Ærø. The e-ferry can carry 30 vehicles and two hundred passengers and is powered by a battery of 4.3 MWh (5800 bhp). The vessel can sail up to 22 nautical miles (25 mi; 41 km) between charges – seven times further than previously possible for an e-ferry. The EU, which supported the project, has stated it aims to roll out 100 or more of these ferries by 2030.

Chapter 18

Ocean liners

Ocean liners are a type of passenger ship primarily used as a form of transportation across seas or oceans. Liners may also carry cargo or mail and may sometimes be used for other purposes (such as for pleasure cruises or as hospital ships). Cargo vessels running to a schedule are sometimes called liners. The category does not include ferries or other vessels engaged in short-sea trading, nor dedicated cruise ships where the voyage itself, and not transportation, is the prime purpose of the trip. Nor does it include tramp steamers, even those equipped to manage limited numbers of passengers. Some shipping companies refer to themselves as 'lines' and their container ships, which often operate over set routes according to established schedules, as 'liners'. Ocean liners are usually strongly built with a high freeboard to withstand rough seas and adverse conditions encountered in the open ocean. Additionally, they are often designed with thicker hull plating than is found on cruise ships and have large capacities for fuel, food, and other consumables on long voyages. The first ocean liners were built in the mid-19th century. Technological innovations such as the steam engine and steel hull allowed larger and faster liners to be built, giving rise to competition between world powers of the time, especially between the United Kingdom and Germany. Once the dominant form of travel between continents, ocean liners were rendered obsolete by the emergence of long-distance aircraft after World War II. Advances in automobile and railway technology also played a role in diminishing their presence. After the retirement of the *Queen Elizabeth 2* in 2008, the only ship still in service as an ocean liner is Cunard's *RMS Queen Mary 2*.

Ocean liners were the primary mode of intercontinental travel for over a century, from the mid-19th century until they began to be supplanted by airliners in the 1950s. In addition to passengers, liners carried mail and cargo. Ships contracted to carry British Royal Mail used the designation RMS. Liners were also the preferred way to move gold and other high-value cargoes. The busiest route for liners was on the North Atlantic with ships travelling between Europe and North America. It was on this route that the fastest, largest, and most advanced liners travelled, though most ocean liners historically were mid-sized vessels, which served as the common carriers of passengers and freight between nations and among mother countries and

DOI: 10.1201/9781003342366-22

their colonies and dependencies in the pre-jet age. Such routes included Europe to African and Asian colonies, Europe to South America, and migrant traffic from Europe to North America in the 19th and first two decades of the 20th centuries, and to Canada and Australia after World War II.

Shipping lines are companies engaged in shipping passengers and cargo, often on established routes and schedules. Regular scheduled voyages on a set route are called 'line voyages' and vessels (passenger or cargo) trading on these routes to a timetable are called 'liners'. The alternative to liner trade is 'tramping' whereby vessels are notified on an ad hoc basis as to the availability of a cargo to be transported. (In older usage, 'liner' also referred to ships of the line, that is, line-of-battle ships, but that usage is now rare.) The term 'ocean liner' has come to be used interchangeably with 'passenger liner', although it can refer to a cargo liner or cargo-passenger liner. The advent of the Jet Age and the decline in transoceanic ship service brought about a gradual transition from passenger ships to modern cruise ships as a means of transportation. For ocean liners to remain profitable, cruise lines modified some of them to operate on cruise routes, such as the *SS France*. Certain characteristics of older ocean liners made them unsuitable for cruising, such as high fuel consumption, deep draught preventing them from entering shallow ports, and cabins (often windowless) designed to maximise passenger numbers rather than comfort. The Italian Line's *SS Michelangelo* and *SS Raffaello*, the last ocean liners to be built primarily for crossing the North Atlantic, could not be converted economically and had short careers (Figure 18.1).

Figure 18.1 RMS *Queen Mary 2* transiting the Suez Canal, Egypt.

DEVELOPMENT OF THE OCEAN LINER

At the beginning of the 19th century, the Industrial Revolution and the inter-continental trade rendered the development of secure links between continents imperative. Being at the top among the colonial powers, Great Britain needed stable maritime routes to connect various parts of the British Empire: the Far East, India, Australia, etc. The birth of the concept of international water and the lack of any claim to it simplified navigation. In 1818, the Black Ball Line, with a fleet of sailing ships, offered the first regular passenger service with an emphasis on passenger comfort, from England to the USA. In 1807, Robert Fulton succeeded in applying steam engines to ships. He built the first ship that was powered by this technology, the *Clermont*, which succeeded in travelling between New York City and Albany, New York, in 30 hours before entering regular service between the two cities. Soon after, other vessels were built using this same innovation. In 1816, the *Élise* became the first steamship to cross the English Channel. Another important advance came in 1819 when the *SS Savannah* became the first steamship to cross the Atlantic Ocean. She left the city of Savanna, Georgia, and arrived in Liverpool, England, in 27 days. Most of the distance was covered by sailing; steam power was not used for more than 72 hours during the passage. The public enthusiasm for the innovative technology was not high, as none of the 32 people who had booked a seat boarded the ship for that historic voyage. Although the *SS Savannah* had proven that a steamship was capable of crossing oceans, the public were not yet prepared to trust such means of travel on the open sea, and, in 1820, the steam engine was removed from the ship. Work on this technology continued and a new step was taken in 1833. *Royal William* managed to cross the Atlantic by using steam power for the entire voyage. The sail was used only when the boilers were cleaned. There were still many sceptics, however, and in 1836, the Irish scientific writer Dionysius Lardner declared that 'as the project of making the voyage directly from New York to Liverpool, it was perfectly chimerical, and they might as well talk of making the voyage from New York to the moon'. The last step towards long-distance travel using steam power was taken in 1837 when *SS Sirius* left Liverpool on 4 April and arrived in New York 18 days later, that is 22 April, after a turbulent crossing. Too little coal was prepared for the crossing, and the crew had to burn cabin furniture to complete the voyage. The journey took place at a speed of 8.03 kn (9.2 mph; 14.87 km/h). The voyage was made possible using a condenser, which fed the boilers with fresh water, avoiding having to periodically shut down the boilers to remove the salt. However, the feat was short-lived. The next day, *SS Great Western*, designed by railway engineer Isambard Kingdom Brunel, arrived in New York. She left Liverpool on 8 April and overtook *Sirius's* record with an average speed of 8.66 kn (9.96 mph; 16.03 km/h). The race for speed had commenced, and, with it, the tradition of the Blue Riband.

With the *SS Great Western*, Isambard Kingdom Brunel laid the foundations for new shipbuilding techniques. He realised that the carrying capacity of a ship increases with the cube of its dimensions, whilst the water resistance only increases as the square of its dimensions. This means that large ships are more fuel-efficient, something especially important for long voyages across the Atlantic. Constructing large ships was therefore more profitable. Moreover, migration to the Americas increased enormously. These movements of population were a financial windfall for the shipping companies, some of the largest of which were founded during this time, including P&O of the United Kingdom in 1822 and the Compagnie Générale Transatlantique of France in 1855. The steam engine also allowed ships to provide regular service without the use of sails. This aspect particularly appealed to the postal companies, which leased the services of ships to serve clients separated by the ocean. *RMS Umbria* and her sister ship *RMS Etruria* were the last two Cunard liners of the period to be fitted with auxiliary sails. Both ships were built by John Elder & Co. of Glasgow, Scotland, in 1884. They were record breakers by the standards of the time, and were the largest liners then in service, plying the Liverpool to New York route.

In 1839, Samuel Cunard founded the Cunard Line and became the first to dedicate the activity of his shipping company to the transport of mail, thus ensuring regular services on a given schedule. The company's vessels operated the routes between the United Kingdom and the United States. In 1840, Cunard Line's *RMS Britannia* began its first regular passenger and cargo service by steamship, sailing from Liverpool to Boston that year. As the size of ships increased, the wooden hull became increasingly fragile. Beginning with the use of an iron hull in 1845, steel hulls eventually solved this problem. The first ship to be both iron-hulled and equipped with a screw propeller was the *SS Great Britain*, a creation of Brunel. Her career was, however, disastrous and short. She was run aground and stranded at Dundrum Bay in Northern Ireland, in 1846. In 1884, she was retired to the Falkland Islands where she was used as a warehouse, quarantine ship, and coal hulk until she was scuttled in 1937. The American company Collins Line equipped its ships with cold rooms, heating systems, and various other innovations, but the operation proved prohibitively expensive. The sinking of two of its ships was a further major blow to the company, which was dissolved in 1858. In the same year, Brunel built his third and last giant, the *SS Great Eastern*. The ship was, for 43 years, the largest passenger ship ever built with a capacity to carry 4,000 passengers. Her career was marked by a series of failures and incidents, one of which was an explosion on board during her maiden voyage.

Many ships owned by German companies like Hamburg America Line and Norddeutscher Lloyd were sailing from major German ports, such as Hamburg and Bremen, to the USA during this time. In 1858, the *SS Austria* suffered a catastrophic fire off the coast of Newfoundland. The ship, built in Greenock and sailing between Hamburg and New York twice a month, sank with the loss of all but 89 of the 542 passengers on board. In the British market, Cunard Line and White Star Line (the latter after being bought by Thomas Ismay in 1868) competed strongly against each other in the late 1860s. The struggle was symbolised by the attainment of the Blue Riband, which the two companies achieved several times around the end of the century. The luxury and technology of ships were also evolving. Auxiliary sails became obsolete and disappeared completely at the end of the century. Possible military use of passenger ships was envisaged and, in 1889, *RMS Teutonic* became the first auxiliary cruiser in history. In the time of war, these ships could be easily equipped with cannons and used in cases of conflict. In 1870, the White Star Line's *RMS Oceanic* set a new standard for ocean travel by having its first-class cabins amidships, with the added amenity of large portholes, electricity, and running water. The size of ocean liners increased from 1880 to meet the needs of immigration to the USA and Australia. The *SS Ophir* was a 6,814 tonne steamship owned by the Orient Steamship Co. and was fitted with refrigeration equipment. She plied the Suez Canal route from England to Australia during the 1890s, up until the years leading to World War I when she was converted into an armed merchant cruiser. In 1897, Norddeutscher Lloyd launched the *SS Kaiser Wilhelm der Grosse*. She was followed three years later by three sister ships. The ship was both luxurious and fast, managing to steal the Blue Riband from the British. She was also the first of the 14 ocean liners with four funnels that emerged around that time. The ship needed only two funnels, but more funnels gave passengers a feeling of safety and power. In 1900, the Hamburg America Line competed with its own four-funnel liner, *SS Deutschland*. She quickly obtained the Blue Riband for the company. This race for speed, however, was a detriment to passengers' comfort and generated strong vibrations, which made her owner lose any interest in her after she lost the Blue Riband to another ship owned by Norddeutscher Lloyd. She was only used for 10 years for transatlantic crossing before being converted into a cruise ship. From 1900 to 1907, the Blue Riband remained in German hands.

In 1902, the American businessman J.P. Morgan embraced the idea of a maritime empire comprising many companies. He founded the International Mercantile Marine Co., a trust that originally comprised only American shipping companies. The trust then absorbed the British company's Leyland Line and White Star Line. Seeing this as a risk to British maritime supremacy, the British government decided to intervene to regain the ascendancy. Although German liners dominated the market in terms of speed, British liners dominated in terms of size. *RMS Oceanic* and the Big Four of the White Star Line were the first liners to surpass *SS Great Eastern* as the largest passenger ships afloat. Their owner was American (as mentioned above, White Star Line had been absorbed into J.P. Morgan's trust; however, faced with this major competition, the British government contributed financially to Cunard Line's construction of two liners of unmatched size and speed, under the condition that they be available for conversion into armed cruisers if and needed by the Royal Navy). The result of this partnership was the completion in 1907 of two sister ships: the *RMS Lusitania* and *RMS Mauretania*, both of which won the Blue Riband during their respective maiden voyages. The latter retained this distinction for 20 years. Their great speed was achieved using turbines instead of conventional expansion machines. In response to the competition from Cunard Line, White Star Line ordered the *Olympic-class* liners at the end of 1907. The first of these three liners, *RMS Olympic*, completed in 1911, had a fine career, although punctuated by incidents. This was not the case for her sister, the *RMS Titanic*, which sank on her maiden voyage on 15 April 1912, resulting in several changes to maritime safety practices. As for the third sister *HMS Britannic*, she never served her intended purpose as a passenger ship, as she was drafted in World War I as a Hospital Ship and sank after hitting a sea mine in 1916.

At the same time, France tried to mark its presence with the completion in 1912 of the *SS France*, which was owned by the Compagnie Générale Transatlantique. Germany soon responded to the competition from the British. From 1912 to 1914, Hamburg America Line completed a trio of liners significantly larger than the White Star Line's Olympic-class ships. The first to be completed, in 1913, was *SS Imperator*. She was followed by *SS Vaterland* in 1914. The construction of the third liner, *SS Bismarck*, was temporarily paused by the outbreak of World War I. Sadly, World War I was a challenging time for the liners. Some of them, like the *SS Mauretania, SS Aquitania*, and *SS Britannic*, were transformed into hospital ships during the conflict. Others became troop transports, while some, such as the *SS Kaiser Wilhelm der Grosse*, participated as warships. Troop transportation was extremely popular due to the liners' enormous size. Liners converted into troop ships were painted in dazzle camouflage to reduce the risk of being torpedoed by enemy submarines. Despite this, the war marked the loss of many liners. *SS Britannic*, whilst serving as a hospital ship, sank in the Aegean Sea in 1916 after she struck a mine. Numerous incidents of torpedoing took place and large numbers of ships were sank. *SS Kaiser Wilhelm der Grosse* was defeated and scuttled after a fierce battle with *HMS Highflyer* off the coast of west Africa, while her sister ship *SS Kronprinz Wilhelm* served as a commerce raider. The torpedoing and sinking of *RMS Lusitania* on 7 May 1915 caused the loss of 128 American lives at a time when the United States was still neutral. Although other factors came into play, the loss of American lives in the sinking strongly pushed the USA to favour the Allied Powers and facilitated America's entry into the war. The losses sustained by the Allied Powers was compensated through the Treaty of Versailles in 1919. This led to the awarding of many German liners to the Allies. The Hamburg America Line's trio (*SS Imperator*, *SS Vaterland*, and *SS Bismarck*) was divided between the Cunard Line, White Star Line, and the United States Lines, while the three surviving ships of the *Kaiser class* were requisitioned for use by the US Navy. The *Tirpitz*, whose construction was delayed by the outbreak of war, eventually became the *RMS Empress of*

Australia. Of the German superliners, only the *SS Deutschland* avoided this fate on account of her poor condition.

After a period of reconstruction, the shipping companies recovered quickly from the damage caused by the First World War. The ships, whose construction was started before the war, such as *SS Paris* of the French Line, were completed and put into service. Prominent British liners, such as the *SS Olympic* and the *SS Mauretania*, were also put back into service and had a successful career in the early 1920s. More modern liners were also built, such as *SS Île de France* (completed in 1927). The United States Lines, having received the *SS Vaterland*, renamed her *Leviathan* and made her the flagship of the company's fleet. Because all USA registered ships counted as an extension of the United States territory, the National Prohibition Act made American liners alcohol-free, causing alcohol-seeking passengers to choose other liners for travel and reducing profits for the United States Lines. In 1929, Germany returned to the scene with the two ships of Norddeutscher Lloyd, *SS Bremen* and *SS Europa*. Bremen won the Blue Riband from Britain's *SS Mauretania* after the latter had held it for 20 years. Soon, Italy also entered the scene. The Italian Line completed *SS Rex* and *SS Conte di Savoia* in 1932, breaking the records of both luxury and speed (*SS Rex* won the westbound Blue Riband in 1933). France re-entered with the *SS Normandie* of the French Line Compagnie Générale Transatlantique. The ship was the largest ship afloat at the time of her completion in 1932. She was also the fastest, winning the Blue Riband in 1935.

A crisis arose when the USA drastically reduced its immigrant quotas, causing shipping companies to lose a large part of their income forcing them to adapt to new revenue streams. The Great Depression also played a significant role, causing a drastic decrease in the number of people crossing the Atlantic and at the same time reducing the number of profitable transatlantic voyages. In response, shipping companies redirected many of their liners to the more profitable cruise service. In 1934, in the UK, Cunard Line and White Star Line were seriously financially challenged. The Chancellor of the Exchequer, Neville Chamberlain, proposed to merge the two companies to solve their financial problems. The merger took place in 1934 and launched the construction of the *RMS Queen Mary* while progressively sending their older ships to be scrapped. The *RMS Queen Mary* was the fastest ship of her time and the largest for a brief period, capturing the Blue Riband twice off the *SS Normandie*. The construction of a second ship, the *RMS Queen Elizabeth*, was interrupted by the outbreak of World War II. From the start of the conflict, German liners were requisitioned, and many were turned into barracks ships. It was during this activity that the *SS Bremen* caught fire and was scrapped in 1941. During the war, *RMS Queen Elizabeth* and *RMS Queen Mary* provided distinguished service as troopships. Many liners were sunk causing great loss of life; in World War II, the three worst disasters involved the loss of the *Cunarder Lancastria* in 1940 off Saint-Nazaire to German bombing while attempting to evacuate troops of the British Expeditionary Force from France, with the loss of more than 3000 lives; the sinking of the *Wilhelm Gustloff*, after the ship was torpedoed by a Soviet submarine, with more than 9000 lives lost, making it the deadliest maritime disaster in history; and the sinking of *SS Cap Arcona* with more than 7000 lives lost, both in the Baltic Sea, in 1945. *SS Rex* was bombarded and sunk in 1942, while *SS Normandie* caught fire, capsized, and sank in New York in 1942 while being converted for troop duty. The *SS Empress of Britain* was attacked by German planes, then torpedoed by a U-boat, when tugs tried to tow her to safety on 28 October 1940.

After the war, some ships were again transferred from the defeated nations to the winning nations as war reparations. This was the case of *the SS Europa*, which was ceded to France and renamed *Liberté*. The US Government was extremely impressed with the service of the Cunard's *RMS Queen Mary* and *RMS Queen Elizabeth* as troopships during the war. To ensure reliable and fast troop transport in case of a war against the Soviet Union, the US Government sponsored the construction of the *SS United States* and entered it into service for United States Lines in 1952. She won the Blue Riband on her maiden voyage in that year and held it until Richard Branson won it back in 1986 with *Virgin Atlantic Challenger II*. One year later, in 1953, Italy completed the *SS Andrea Doria*, which later sank in 1956 after a collision with *MS Stockholm*.

Before World War II, aircraft had not posed a significant economic threat to ocean liners. Most pre-war aircraft were noisy, vulnerable to severe weather, few had the range needed for transoceanic flights, and all were expensive and had a small passenger capacity. The war accelerated the development of large, long-ranged aircraft. Four-engine bombers such as the Avro Lancaster and Boeing B-29 Superfortress, with their range and massive carrying capacity, were natural prototypes for post-war next-generation airliners. Jet engine technology also accelerated due to the wartime development of jet aircraft. In 1953, the De Havilland Comet became the first commercial jet airliner; the Sud Aviation Caravelle, Boeing 707, and Douglas DC-8 followed, and much long-distance travel was done by air. The Italian Line's *SS Michelangelo* and *SS Raffaello*, launched in 1962 and 1963, respectively, were two of the last ocean liners to be built primarily for liner service across the North Atlantic. Cunard's transatlantic liner, *RMS Queen Elizabeth 2*, was also used as a cruise ship rather than as an ocean liner. By the early 1960s, 95% of passenger traffic across the North Atlantic was by aircraft. Thus, the reign of the ocean liners ended. By the early 1970s, many passenger ships continued their service in the cruising industry. In 1982, during the Falklands War, three active or former liners were requisitioned for war service by the British Government. The liners *RMS Queen Elizabeth 2* and *SS Canberra*, were requisitioned from Cunard and P&O to serve as troopships, carrying British Army personnel to Ascension Island and the Falkland Islands to recover the Falklands from invading Argentine forces. The P&O educational cruise ship and former British India Steam Navigation Company liner *Uganda* was requisitioned as a hospital ship and served after the war as a troopship until the RAF Mount Pleasant station was built at Stanley, which could manage trooping flights. By the first decade of the 21st century, only a few former ocean liners were still in existence, some like *SS Norway*, were sailing as cruise ships while others, like the *RMS Queen Mary*, were preserved as museums, or laid up at pier side like the *SS United States*. After the retirement of the *RMS Queen Elizabeth 2* in 2008, the only ocean liner in service was the *RMS Queen Mary 2*, built in 2003–2004, used for both point-to-point line voyages and for cruises.

Four ocean liners that were made before World War II survive today, as they have been preserved as museums and hotels. The Japanese ocean liner *Hikawa Maru* (1929) has been preserved in Naka-ku, Yokohama, Japan, as a museum ship, since 1961. *RMS Queen Mary* (1934) was preserved in 1967 after her retirement and became a museum and floating hotel in Long Beach, California. In the 1970s, *SS Great Britain* (1843) was preserved, and now resides in Bristol, England, as another museum. The latest ship to undergo preservation is *MV Doulos* (1914), which became a dry berthed luxury hotel on Bintan Island, Indonesia. Post-war ocean liners still extant are the *SS United States* (1952), docked in Philadelphia since 1996; the *SS Rotterdam* (1958), moored in Rotterdam as a museum and hotel since 2008; and the *RMS Queen Elizabeth 2* (1967), a floating luxury hotel and museum at Mina Rashid, Dubai, since 2018. *MV Astoria* (1948) (originally *MS Stockholm*, which collided with *Andrea Doria* in 1956) was in active service for Cruise & Maritime Voyages until operations ceased in 2020 due to the COVID-19 pandemic. In July 2021, the ship was purchased at the line's bankruptcy auction by a US-based investor group who intended to use her for island cruises from Lisbon. In August of that year, it was repurchased by Brock Pierce to be transformed into a hotel along with *MV Funchal*. In 2021, *MV Funchal* (1961) was purchased by Brock Pierce for over US$1 million. Originally operated by the Portuguese shipping company Empresa Insulana de Navegação, it is currently being renovated into a hotel.

CHARACTERISTICS

Size and speed

Since their beginning in the 19th century, ocean liners have had to meet growing demands. The first liners were small and overcrowded, leading to unsanitary conditions on board. Eliminating these problems required larger ships, to reduce the crowding of passengers, and faster ships, to reduce the duration of transatlantic crossings. The development of iron and steel hulls and steam power allowed for these technological advances. The ship's *SS Great Western* (1340 tonnes) and *SS Great Eastern* (18,915 tonnes) were constructed in 1838 and 1858, respectively. The record set by the *SS Great Eastern* was not beaten until 43 years later in 1901 when *RMS Celtic* (20,904 tonnes) was completed. Tonnage then grew profoundly: the first liners to have a tonnage that exceeded 20,000 were the Big Four of the White Star Line. The Olympic class ocean liners, first completed in 1911, were the first to have a tonnage that exceeded 45,000. *SS Normandie*, completed in 1935, had a tonnage of 79,280. In 1940, *RMS Queen Elizabeth* raised the record of size to a tonnage of 83,673. She was the largest passenger ship ever constructed until 1997. In 2003, *RMS Queen Mary 2* became the largest, at 149,215 metric-tonnes.

In the early 1840s, the average speed of liners was less than 10 knots (11 mph; 18 kp/h), with a crossing of the North Atlantic taking about 12 days. In the 1870s, the average speed of liners increased to around 15 knots (17 mph; 27 km/h) with the duration of a transatlantic crossing shortened to around 7 days. This owed to the technological progress made in the propulsion of ships: rudimentary steam boilers gave rise to more elaborate machineries and the paddlewheel gradually disappeared altogether, replaced first by one helix and then by two helix propellers. At the beginning of the 20th century, Cunard Line's *RMS Lusitania* and *RMS Mauretania* reached a speed of 27 knots (31 mph; 50 km/h). Their records seemed unbeatable, and most shipping companies abandoned the race for speed in favour of size, luxury, and safety. The advent of ships with diesel engines, and of those whose engines were oil-burning, such as the *SS Bremen*, in the early 1930s, relaunched the race for the Blue Riband. The *SS Normandie* won it in 1935 before being snatched by *RMS Queen Mary* in 1938. It was not until 1952 that *SS United States* set a record that remains today: 34.5 knots (39.7 mph; 63 km/h) taking only 3 days and 12 hours to cross the Atlantic from New York to Southampton.

Passenger cabins and amenities

The first ocean liners were designed to carry mostly migrants. On board sanitary conditions were often deplorable and epidemics were frequent. In 1848, maritime laws imposing hygiene rules were adopted, which drastically improved the on-board living conditions. Gradually, two distinct classes developed: the cabin class and the steerage class. The passengers travelling on the former were wealthy passengers and they enjoyed certain comforts associated with that class. The passengers travelling in steerage were members of the middle or working classes. In that class, they were packed in large dormitories. Until the beginning of the 20th century, they did not always have bedsheets and meals. An intermediate class for tourists and members of the middle class gradually appeared. The cabins were then divided into three classes. The facilities offered to passengers developed over time. In the 1870s, the installation of bathtubs and oil lamps caused a sensation on board *RMS Oceanic*. In the following years, the number of amenities became numerous, for example smoking rooms, lounges, and the promenade deck. In 1907, *RMS Adriatic* even offered Turkish baths and a swimming pool. In the 1920s, *SS Paris* was the first liner to provide a movie theatre (Figure 18.2).

Figure 18.2 Vasco da Gama at the Liverpool Cruise terminal, England.

SHIP BUILDERS

The British and Germans were the most famed in shipbuilding during the great era of ocean liners. In Ireland, Harland & Wolff shipyard in Belfast were particularly innovative and succeeded in winning the trust of many shipping companies, such as White Star Line. These gigantic shipyards employed a substantial portion of the local population and built hulls, machines, furniture, and lifeboats. Among the other well-known British shipyards were Swan, Hunter, and Wigham Richardson, the builder of *RMS Mauretania*, and John Brown & Company, the builder of *RMS Lusitania*. Germany had many shipyards on the coast of the North Sea and the Baltic Sea, including Blohm & Voss and AG Vulcan Stettin. Many of these shipyards were destroyed during World War II, though a few managed to recover and continue building ships. In France, major shipyards included Chantiers de Penhoët in Saint-Nazaire, made famous for building the *SS Normandie*. This shipyard merged with Ateliers et Chantiers de la Loire shipyard to form the Chantiers de l'Atlantique shipyard, which has built ships including the *RMS Queen Mary 2*.

SHIPPING COMPANIES

Although there were many British shipping companies, two were particularly distinguished: Cunard Line and White Star Line. Both were founded during the 1830s and were engaged in intense competition against one another, possessing the largest and fastest liners in the world until the mid-20th century. It was not until 1934 that financial difficulty caused the two to merge, forming Cunard White Star Ltd. P&O also occupied a large part of the liner business. The Royal Mail Steam Packet Company operated as a state-owned enterprise with its close relationship with the government. Over the course of its history, it took over many shipping companies, becoming one of the largest companies in the world before legal problems led to its liquidation in 1931. On the continent, two rival companies, Hamburg America Line (often referred to as 'HAPAG') and Norddeutscher Lloyd, competed in Germany. The First and Second World Wars dealt much damage to the two companies, both forced to renounce their ships to the winning side in both wars. The two merged to form the modern-day shipping major, Hapag-Lloyd, in 1970. The ocean liner industry in France also consisted of two rival companies: the Compagnie Générale Transatlantique (commonly known as 'Transat' or 'French Line') and Messageries Maritimes. The Compagnie Générale Transatlantique operated on the North Atlantic route with well-known liners such as *SS Normandie* and *SS France*, while Messageries Maritimes operated throughout the French colonies in Asia and Africa. Decolonisation in the second half of the 20th century led to a sharp decline in profit for Messageries Maritimes, and it merged with Compagnie Générale Transatlantique in 1975 to form the Compagnie Générale Maritime. The Netherlands had three main companies: Holland America Line, which operated mostly on the North Atlantic route and with well-known ships like the *SS Nieuw Amsterdam* and *SS Rotterdam*. Unlike the French and German industry, the Holland America Line had no domestic rivals and thus only had to compete with foreign lines. The other two Dutch lines were the Stoomvaart Maatschappij Nederland (SMN), otherwise known as the Netherland Line and the Koninklijke Rotterdamsche Lloyd (KRL). Both offered regular service between the Netherlands and the Dutch East Indies, the Dutch colony in Southeast Asia now known as Indonesia. In Italy, the Italian Line was founded in 1932 because of a merger of three companies. It was known for operating liners such as *SS Rex* and *SS Andrea Doria*. In the US, the United States Lines tried to impose itself on the international scene but failed to compete with its European rivals. The Japanese established Nippon Yusen, also known as NYK Lines, which ran trans-Pacific liners such as the *Hikawa Maru* and the *Asama Maru*.

ROUTES

North Atlantic

The most important of all routes taken by ocean liners was the North Atlantic route. It accounted for a large part of the clientele, who travelled between the ports of Liverpool, Southampton, Hamburg, Le Havre, Cherbourg, Cobh, and New York City. The profitability of this route came from European migration to North America. The need for speed influenced the construction of liners for this route, and the Blue Riband was awarded to the liner with the highest speed. The route was not without danger, however, as storm and icebergs were (and continue to be) common in the North Atlantic. Many shipwrecks occurred on this route, among them the *RMS Titanic*.

South Atlantic

The South Atlantic was the route frequented by liners bound for South America, Africa, and sometimes even Oceania. The White Star Line had some of its ships, such as the *Suevic*, on the Liverpool-Cape Town-Sydney route. Given there was little competition in the South Atlantic as there was in the North Atlantic, the incidence of shipwrecks was less. The German shipping company, Hamburg Süd, is the most famous operator of this route, with the famed *SS Cap Arcona*.

Mediterranean

The Mediterranean Sea was frequented by many ocean liners. Many companies benefited from migration from Italy and the Balkans to the US. Cunard's *RMS Carpathia* served on the Gibraltar-Genoa-Trieste route. Similarly, Italian liners crossed the Mediterranean Sea before entering the North Atlantic Ocean. The construction of the Suez Canal opened the Mediterranean as a popular route to Asia.

Indian Ocean and the Far East

Colonisation made Asia particularly attractive to shipping companies. As early as the 1840s, P&O organised trips to Calcutta via the Suez Isthmus, as the canal had not yet been built. The time it took to travel on this route to India, Southeast Asia, and Japan was long, with many stopovers. The Messageries Maritimes operated on this route, notably in the 1930s, with its motor ships. Similarly, the *La Marseillaise*, put into service in 1949, was one of the flagships of its fleet. Decolonisation in the 1960s and 1970s caused many of these routes to flounder and eventually cease altogether.

Other

The construction of ocean liners was, in some respects, a result of nationalism. The revival of power of the German Navy stemmed from the clear affirmation of Kaiser Wilhelm II to see his country become a major sea power, capable of rivalling Britain. Thus, the *SS Deutschland* (1900) had the honour of becoming the nation's name bearer, an honour she lost after

10 years of a disappointing career. *RMS Lusitania* and *RMS Mauretania* (1907) were built with the help of the British Government with the desire that the United Kingdom would regain its prestige as the world's preeminent sea power. The *SS United States* (1952) was the result of a desire by the US government to possess a large and fast ship that was convertible into a troop transport. The *SS Rex* and *SS Conte di Savoia* (1932) were constructed at the demands of Benito Mussolini. Finally, the construction of *SS France* in 1961 was a result of Charles de Gaulle's desire to build on French national pride and was financed by the French Government. Some liners did gain great popularity. *SS Mauretania* and *SS Olympic* had many admirers during their careers, and their retirement and scrapping caused certain sadness. The same was also true of *Île de France*, whose scrapping aroused strong emotion from her admirers.

MARITIME DISASTERS AND INCIDENTS

Some ocean liners are known today because of their sinking with great loss of lives. In 1873, *RMS Atlantic* struck an underwater rock and sank off the coast of Nova Scotia, Canada, killing at least 535 people. In 1912, the sinking of the *RMS Titanic*, which took approximately 1,500 lives, highlighted the overconfidence of the shipping companies in their ships, such as the failure to have sufficient lifeboats on board. Safety measures at sea were re-examined following the incident. Two years later, in 1914, *RMS Empress of Ireland* sank in the St. Lawrence River after colliding with another ship. A total of 1,012 people lost their lives. Among the other sinkings are the torpedo sinking of the *RMS Lusitania* in 1915, which resulted in the loss of 1198 lives and provoked an international outcry, the naval mine sinking of the *HMHS Britannic* in 1916, and that of *MS Georges Philippar*, which caught fire and sank in the Gulf of Aden in 1932, killing 54 people. In 1956, the sinking of *SS Andrea Doria*, with the loss of 46 lives, after a collision with *MS Stockholm* made the headlines throughout Europe. In 1985, the Italian liner *MS Achille Lauro* was hijacked off the coast of Egypt by members of the Palestinian Liberation Front, resulting in the death of one of the hostages. In 1994, she caught fire and sank off the coast of Somalia.

This completes Part III. In Part IV, we will turn our attention towards the various types of vessels used in the offshore supply and support sector.

Part IV

Construction and support vessels

Chapter 19

Cable layers

A cable layer or cable ship is a deep-sea vessel designed and used to lay underwater cables for telecommunications, electric power transmission, military, or other purposes. Cable ships are distinguished by large cable sheaves for guiding cable over the bow or stern or both. Bow sheaves, some exceptionally large, were characteristic of all cable ships in the past, but newer ships are tending towards having stern sheaves only. The names of cable ships are often preceded by 'CS' as in *CS Long Lines*. The first transatlantic telegraph cable was laid by cable-layers in 1857–1858. It briefly enabled telecommunication between Europe and North America before misuse resulted in the failure of the line. In 1866, the *SS Great Eastern* successfully laid two transatlantic cables, securing future communication between the continents. Cable layers have unique requirements related to having long idle periods in port between cable laying or repairs, operation at low speeds or stopped at sea during cable operations, extended periods running astern (less frequent as stern layers are now common), high manoeuverability, and a fair speed to reach operation areas. As such, modern cable layers differ from their predecessors. There are two main types of cable-layers: cable repair ships and cable-laying ships. Cable repair ships, like the Japanese *Tsugaru Maru*, tend to be smaller and more manoeuverable; they are capable of laying cable, but their primary job is fixing or repairing broken sections of cable. A cable-laying ship, like the *CS Long Lines*, is designed to lay new cables. Such ships are bigger than repair ships and less manoeuverable; their cable storage drums are also larger and are set in parallel so one drum can feed into another, allowing them to lay cable much faster. These ships are also generally equipped with a linear cable engine (LCE) that helps them lay cable quickly. By locating the manufacturing plant near a harbour, the cable can be loaded into the ship's hold as it is being manufactured (Figure 19.1).

The newest design of cable layers, though, is a combination of cable-laying and repair ships. An example is the *USNS Zeus* (T-ARC-7), the only US naval cable layer-repair ship. *USNS Zeus* uses two diesel-electric engines that produce 5,000 horsepower each and can carry her up to 15 kts (17 mph; 27 km/h). She can lay about 1,000 miles (1,600 km) of telecommunications cable to a depth of 2,700 m (9,000 ft). The purpose of *USNS Zeus*

DOI: 10.1201/9781003342366-24

Figure 19.1 Cable ship *Ile de Bréhat.*

was to be a cable ship that could do anything required of her, so the ship was built to be able to lay and retrieve cable from either the bow or the stern with ease. This design was like that of the first cable ship, the *SS Great Eastern*. To ensure that the cable is laid and retrieved properly, specially designed equipment must be used. Different equipment is employed on cable-laying ships depending on what the specific job entails. To retrieve damaged or mislaid cable, a grapple system is used to gather the cable from the ocean floor. There are several types of grapples, each with certain pros and cons. The grapples are attached to the vessel via a grapple rope. Originally, these were manufactured from a mix of steel and manila lines but are now made from synthetic materials. This ensures that the line is strong yet can flex and strain under the weight of the grapple. The line is pulled up by reversing the LCE. The LCE is used to feed the cable down to the ocean floor; however, this can also be reversed and used to bring the cable back up in the event it requires repair. The engines can feed 244 m (800 ft) of cable every minute. However, cable ships are usually limited to 8 kts (9.2 mi; 14.8 km/h) while laying cable to ensure the cable lies on the sea floor properly, and to compensate for any small adjustments in course that might affect the cables' position. This is important as the cable needs to be located for preventative and corrective maintenance. LCEs are also equipped with a brake system that allows the flow of cable to be controlled or stopped should problem arise. A common system often found on modern vessels is the fleeting drum. This is a mechanical drum, which is fitted with eoduldes (raised surfaces on the drum face). The eodules help slow and guide the cable into the LCE. Cable ships also use 'ploughs', which are suspended under the vessel. The ploughs use jets of high-pressure water to bury the cable as much as 0.91 m (3 ft) under the sea floor. This prevents fishing vessels from snagging the cables as they thrall their nets.

When coaxial cables were introduced as submarine cables, a new issue with cable-laying was encountered. These cables had periodic repeaters in line with the cable and powered through it. Repeaters overcame significant transmission problems on submarine cables. The difficulty with laying repeaters is that there is a bulge where they are spliced into the cable, and this causes problems passing through the sheave. British ships, such as the *HMTS Monarch* and *HMTS Alert*, solved the problem by providing a trough for the repeater to bypass the sheave. A rope connected in parallel to the repeater went through the sheave which pulled the cable back in to the sheave after the repeater had passed. It was normally necessary for the ship to slow down while the repeater was being laid. American ships, for a time, tried using flexible repeaters which passed through the sheave. However, by the 1960s, they were also using rigid repeaters like the British system. Another issue with coaxial repeaters is that they are much heavier than the cable. To ensure that they sink at the same rate as the cable (which can take some time to reach the bottom) and keep the cable straight, the repeaters are fitted with parachutes.

In this chapter, we have examined the role and function of the oddity that is the cable ship. In the next chapter, we will look at some of the main types of offshore construction support vessels.

Chapter 20

Construction support vessels

PIPELAYING SHIPS

A pipelaying ship is a maritime vessel used in the construction of subsea infrastructure. It serves to connect oil production platforms with refineries on shore. To accomplish this goal, a typical pipelaying vessel carries a heavy lift crane, used to install pumps and valves, and equipment to lay pipe between subsea structures. Lay methods consist of J-lay and S-lay and can be reel-lay or welded length by length. Pipelaying ships make use of dynamic positioning systems or anchor spreads to maintain the correct position and speed while laying pipes. The most recent ships are capable of laying subsea pipework in water depths of more than 2,500 m (8,202 ft). The term pipe laying vessel or pipelayer refers to any vessel capable of laying pipe on the ocean floor. It can also refer to dual activity ships. These vessels are capable of laying pipe on the ocean floor in addition to their primary job. Examples of dual activity pipelayers include barges, modified bulk carriers, modified drillships, semi-immersible laying vessels, and others. The majority of pipelaying vessels are owned and operated by oil and gas companies (Figure 20.1).

HEAVY LIFT AND CRANE VESSELS

A crane vessel, crane ship, or floating crane is a ship with a crane specialised in lifting heavy loads. The largest crane vessels are used for offshore construction. Conventional monohulls are used, but the largest crane vessels are often catamaran or semi-submersible types, as they have increased stability. On a sheerleg crane, the crane is fixed and cannot rotate, and the vessel, therefore, is manoeuvred to place loads. In medieval Europe, crane vessels which could be flexibly deployed in the whole port basin were introduced as early as the 14th century. During the age of sail, the sheer hulk was used extensively as a floating crane for tasks that required heavy lift. At the time, the heaviest single components of ships were the main masts, and sheer hulks were essential for removing and replacing them, but they

DOI: 10.1201/9781003342366-25

Figure 20.1 Pipelayer *Seven Navica*, Leith Docks, Edinburgh, Scotland.

were also used for other purposes. Some crane vessels had engines for propulsion, others needed to be towed with a tugboat. In 1920, the 1898-built battleship *USS Kearsarge* was converted to a crane ship when a crane with a capacity of 250 tonnes was installed. Later renamed *Crane Ship No.1*, she was used, amongst other things, to place guns and other heavy items on other battleships under construction. In 1939, *Crane Ship No.1* was used to raise the sunk submarine, *USS Squalus*. In 1942, the crane ships aka 'Heavy Lift Ships' *SS Empire Elgar* (PQ16), *SS Empire Bard* (PQ15), and *SS Empire Purcell* (PQ16) were sent to the Russian Arctic ports of Archangel, Murmansk, and Molotovsk (since renamed Sererodvinsk). Their role was to enable the unloading of the Arctic convoys where port installations were either destroyed by German bombers or were non-existent (as at Bakaritsa quay in Archangel). In 1949, J. Ray McDermott had *Derrick Barge Four* built, a barge that was outfitted with a revolving crane capable of lifting 150 tonnes. The arrival of this type of vessel changed the direction of the offshore construction industry. Instead of constructing oil platforms in parts, jackets and decks could be built onshore as modules. For use in the shallow part of the Gulf of Mexico, these barges proved extremely useful and resilient (Figure 20.2).

In 1963, Heerema converted a Norwegian tanker, *Sunaas*, into a crane vessel with a capacity of 300 tonnes, the first one in the offshore industry that was 'ship-shaped'. Following the refit, the *Sunaas* was renamed *Global Adventurer* and was the first vessel of its kind to operate in the harsh environment of the North Sea. In 1978, Heerema had two semi-submersible crane vessels built, *Hermod* and *Balder*, each with one 2,000 tonnes and

Figure 20.2 Heavy-lift vessel *Biglift Barentsz* in Schiedam, The Netherlands.

one 3,000 tonne crane. Both were later upgraded to a higher capacity. This type of crane vessel was much less sensitive to sea swell, so it was possible to operate in the North Sea even during the winter months. The high stability also allowed for heavier lifts than was possible with a monohull. The larger capacity of the cranes reduced the installation time of a platform from a whole season to just a few weeks. Inspired by this success, similar vessels were built. In 1985, DB-102 was launched for McDermott, with two cranes with a capacity of 6000 metric tonnes each. Micoperi ordered *M7000* in 1986, which was designed and built with two cranes of 7000 metric tonnes each. Due to the oil glut in the mid-1980s, however, the boom in the offshore industry ended, resulting in a series of collaborations. In 1988, a joint venture between Heerema and McDermott was formed, HeereMac. In 1990, Micoperi was forced to apply for bankruptcy. Saipem, at the beginning of the 1970s, a large heavy lift contractor, but only a small player in this field at the end of the 1980s, acquired *M7000* from Micoperi in 1995, later renaming it *Saipem 7000*. In 1997, Heerema took over *DB-102* from McDermott after the discontinuation of their joint venture. The ship was renamed *Thialf* and, after an upgrade in 2000, both cranes were increased to 7,100 metric tonnes. Up until 2019, *Thialf* was the largest semi-submersible crane vessel in the world, whereafter she was surpassed by the *SSCV Sleipnir*, a US$1 billion vessel built for Heerema (Figure 20.3).

Figure 20.3 *Dockwise Treasure*, off Falmouth, England.

DRILLSHIPS

A drillship is a merchant vessel designed for use in exploratory offshore drilling of new oil and gas wells or for scientific drilling purposes. In most recent years, the vessels are used in deep-water and ultra-deep-water applications and are equipped with the latest and most advanced dynamic positioning systems. The first drillship was the *CUSS I*, designed by Robert F. Bauer of Global Marine in 1955. By 1957, the *CUSS I* had drilled in 121 m (400 ft) deep waters. In 1961, Global Marine started a new drillship era. They ordered several self-propelled drillships each with a rated centreline drilling of 6,096 m (20,000 ft) wells in water depths of 182 m (600 ft). The first was the 1961 *CUSS (Glomar) II*, a 5,500 dwt tonne vessel, costing around US$4.5 million. Built by a Gulf Coast shipyard, the vessel was twice the size of the *CUSS I* and became the world's first purpose-built drillship. In 1962, the Offshore Company elected to build a new type of drillship, larger than that of the Glomar class. This new drillship would feature the first-ever anchor mooring array based on a unique turret system. The vessel was named *Discoverer I*. The *Discoverer I* had no main propulsion engines, meaning she had to be towed out to the drill site (Figure 20.4).

The drillship can be used as a platform to conduct well maintenance or completion work such as casing and tubing installation, subsea tree installations, and well capping. Drillships are often built to the design specifications set by the oil production company and/or investors. From the first drillship *CUSS I* to the *Deepwater Asgard*, the fleet size has been growing ever since. In 2013, the worldwide fleet of drillships topped 80 ships, more than double its size in 2009. Drillships are not only growing but also are becoming capable of performing innovative technology-assisting operations from academic

Figure 20.4 *ENSCO DS-6* drillship in the Saronic Gulf, Greece.

research to ice-breaker class drilling vessels. Drillships are just one way to perform diverse types of offshore drilling. This function can also be performed by semi-submersibles, jack-ups, barges, or platform rigs. Drillships, however, have the functional ability of semi-submersible drilling rigs and have a few unique features that separate them from all others. The first feature is the ship-shaped design. A drillship has greater mobility and can move quickly under its own propulsion from drill site to drill site in contrast to semi-submersibles and jack-up barges and platforms. Drillships can also save time sailing between oilfields worldwide. A typical drillship takes approximately 20 days to transit from the Gulf of Mexico to the Angola offshore basin. Comparatively, a semi-submersible drilling unit must be towed and takes on average 70 days to cover the same distance. Although drillship construction costs are much higher than that of semi-submersibles, their utility and mobility ensure drillship owners can charge higher day rates and get the benefit of lower idle times between assignments. Table 20.1 depicts the industry's way of classifying drill sites into different vintages, depending on their age and water depth:

The drilling operations are incredibly detailed and in-depth. A marine riser is lowered from the drillship to the seabed with a blowout preventer (BOP) at the bottom that connects to the wellhead. The BOP is used to quickly disconnect the riser from the wellhead in times of emergency or in any needed situation. Underneath the derrick is a moonpool, an opening through the hull covered by the rig floor. Some modern drillships have larger derricks that allow dual activity operations, for example simultaneous

Table 20.1 Drill ship vintages

Drillship	Launch date	Water depth (m; ft)
CUSS I	1961	106.6 m (350 ft)
Discoverer 534	1975	2,133 m (7000 ft)
Enterprise	1999	3,048 m (10,000 ft)
Inspiration	2009	3,657 m (12,000 ft)

drilling and casing handling. All drillships have what is called the moon pool. The moon pool is an opening on the base of the hull and depending on the mission the vessel is on, drilling equipment, small submersible crafts and divers may pass through the moon pool. Since the drillship is also a vessel, it can easily relocate to any desired location. But due to their mobility, drillships are not as stable compared to semi-submersible platforms. To maintain their position, drillships often use their anchors or use the ship's computer-controlled system on board to run their dynamic positioning. One of the world's renowned drill ships is Japan's ocean-going drilling vessel the *Chikyū*, which is a research vessel. The *Chikyū* has the remarkable ability to drill a hole of 4 mi (6.4 km) into the seabed, which brings it to a depth of 4.34 mi [7 km (7,000 m or 23,000 ft)] below the seabed. This is more than four times that of any other drillship. In 2011, the Transocean drillship the *Dhirubhai Deepwater KG1* set the world water-depth record at 3,107 m (10,194 ft) whilst conducting directional drilling in India.

DREDGERS

Dredging is the excavation of material from a water environment. Reasons for dredging include improving existing water features; reshaping the land and water features to alter drainage, navigability, and commercial use; constructing dams, dikes, and other controls for streams and shorelines; and recovering valuable mineral deposits or marine life having commercial value. In all but a few situations, the excavation is undertaken by a specialist floating plant, known as a dredger. Dredging is conducted in many separate locations and for many different purposes, but the main objectives are usually to recover material of value or use, or to create a greater depth of water. Dredgers are classified as either suction or mechanical. Dredging has significant environmental impacts. It can disturb marine sediments, leading to both short-term and long-term water pollution, destroy important seabed ecosystems, and can release human-sourced toxins captured in the sediment. Dredging is a four-part process: (1) loosening the material; (2) bringing the material to the surface (together extraction); (3) transportation; and (4) disposal. The extract can be disposed of locally or transported by barge or in a liquid suspension through pipelines. Disposal can be to infill sites, or the material can be used constructively to replenish eroded sand that has been lost to coastal erosion, or constructively to create seawalls, building land, or create whole new landforms such as viable islands in coral atolls. Several ancient authors refer to harbour dredging. The seven arms of the Nile were channelled, with wharfs built at the time of the pyramids (4000 BC). There is evidence to suggest that extensive harbour building occurred in the eastern Mediterranean from around 1000 BC and the disturbed sediment layers give evidence of dredging. At Marseille, dredging phases are recorded from 300 BC onwards, with the most extensive activity occurring during the first century AD. Moreover, the remains of three dredging boats have been uncovered; they were abandoned at the bottom of the harbour during the first and second centuries AD. In modern times, dredging machines were used during the construction of the Suez Canal in the late 1800s, and have continued since to the present day, throughout numerous expansions, and general maintenance. The completion of the Panama Canal in 1914, the most expensive U.S. engineering project at the time, relied extensively on dredging.

Chapter 21

Icebreakers

An icebreaker is a special-purpose ship or boat designed to move and navigate through ice-covered waters and provides safe waterways for other boats and ships. Although the term usually refers to ice-breaking ships, it may also refer to smaller vessels, such as the icebreaking boats that were once used on the canals of the UK. For a ship to be considered an icebreaker, it requires three traits that most normal ships lack: a strengthened hull, an ice-clearing shape, and the power to push through sea ice. Icebreakers work by clearing paths through frozen-over water or pack ice. The bending strength of sea ice is low enough that the ice breaks usually without a noticeable change in the vessel's trim. In cases of very thick ice, an icebreaker can drive its bow onto the ice to break it under the weight of the ship. A build-up of broken ice in front of a ship can slow it down much more than the breaking of the ice itself, so icebreakers have a specially designed hull to direct the broken ice around or under the vessel. The external components of the ship's propulsion system (propellers, propeller shafts, etc.) are at a greater risk of damage than the vessel's hull, so the ability of an icebreaker to propel itself onto the ice, break it, and clear the debris from its path successfully is essential for its safety.

Prior to ocean-going ships, ice-breaking technology was developed on inland canals and rivers using labourers with axes and hooks. The first recorded primitive icebreaker ship was a barge used by the Belgium town of Bruges in 1383 to help clear the town moat. The efforts of the ice-breaking barge were successful enough to warrant the town purchasing four such ships. Ice-breaking barges were continuously used during the colder winters of the Little Ice Age with growing use in the Low Country where a significant amount of trade and transport of people and goods took place. Over the course of the 15th century, the use of ice breakers in Flanders (Oudenaarde, Kortrijk, Leper, Veurne, Diksmuide, and Hulst) was readily established. The use of the ice-breaking barges expanded throughout the 17th century where every town of some importance in the Low Country used some form of icebreaker to keep their waterways clear. Prior to the

DOI: 10.1201/9781003342366-26

17th century, the specifications of icebreakers are unknown. The specifications for ice-breaking vessels show that they were dragged by teams of horses and the heavy weight of the ship was pushed down on the ice breaking it. They were used in conjunction with teams of men with axes and saws, and the technology behind them did not change much until the Industrial Revolution (1760–1840).

Ice-strengthened ships were used in the earliest days of polar exploration. These were originally wooden and based on existing designs, but reinforced, particularly around the waterline with double planking to the hull and strengthening cross members inside the ship. Bands of iron were wrapped around the outside. Sometimes, metal sheeting was placed at the bows, at the stern, and along the keel. Such strengthening was designed to help the ship push through ice and to protect the ship in case it was 'nipped' by the ice. Nipping occurs when ice floes around a ship are pushed against the ship, trapping it as if in a vice, causing damage. This vice-like action is caused by the force of winds and tides acting on the ice formations. The first boats to be used in the polar waters were those of the Indigenous Arctic people. Their kayaks were small human-powered boats with a covered deck, and one or more cockpits, each seating one paddler who stroked a single or double-bladed paddle. Such boats, of course, had no icebreaking capabilities, but they were light and fitted well to be carried over the ice. In the ninth and 10th centuries, the Vikings expanded throughout the north of Europe, eventually North Atlantic, and eventually crossing to Greenland and Svalbard in the Arctic. The Vikings, however, were fortunate in that they operated their ships in waters that were ice-free for most of the year, on account of the conditions of the Medieval Warm Period. In the 11th century, in North Russia, the coasts of the White Sea, named so for being ice-covered for over half of the year, started being settled. The mixed ethnic group of the Karelians and the Russians in the North-Russia that lived on the shores of the Arctic Ocean became known as *Pomors* ('seaside settlers'). Gradually, they developed a special type of small one- or two-mast wooden sailing ships, used for voyages in the icy conditions of the Arctic seas, and later, on the Siberian rivers. These earliest icebreakers were called Kochi. The Kochi's hull was protected by a belt of ice-floe resistant, flush skin-planking along the variable waterline and had a false keel for on-ice portage. If a Koch became squeezed by the ice-fields, its rounded bodylines below the water-line would allow for the ship to be pushed up out of the water and onto the ice with minimal damage.

In the 19th century, similar protective measures were adopted to modern steam-powered icebreakers. Some notable sailing ships at the end of the Age of Sail also featured the egg-shaped form like that of the Pomor boats, for example, the *Fram*, used by Fridtjof Nansen and other Norwegian Polar explorers. *Fram* was the first wooden ship to sail the farthest north (85°57′N) and farthest south (78°41′S). As such, she was one of the strongest wooden ships ever built. An early ship designed to operate in icy conditions was the 51 m (167 ft) wooden paddle steamer, *City Ice Boat No.1*, which was built for the city of Philadelphia by Vandusen & Birelyn in 1837. The ship was powered by two 250 horsepower (190 kW) steam engines and her wooden paddles were reinforced with iron coverings. With a rounded shape and strong metal hull, the *Pilot* of 1864 was an important predecessor of modern icebreakers with propellers. The ship was built on the orders of the merchant and shipbuilder Mikhail Britnev. She had the bow altered to achieve an ice-clearing capability (20 degree raise from the keel line). This allowed the *Pilot* to push herself on the top of the ice and consequently break it. Britnev fashioned the bow of his ship after the shape of old Pomor boats, which had been navigating icy waters of the White Sea and the Barents Sea for centuries. The *Pilot* was used between 1864 and 1890 for navigation in the Gulf of Finland between Kronstadt and Oranienbaum, thus extending the summer navigation season by several weeks. Inspired by the success of *Pilot*, Mikhail Britnev built a second similar vessel *Boy* (English: *Breakage*) in 1875 and a third *Booy* (English: Buoy) in 1889. The freezing winter of 1870–1871 caused the Elbe River and the port of Hamburg to freeze over, causing a prolonged halt to navigation and huge commercial losses. Carl Ferdinand Steinhaus reused the altered bow of the *Pilot's* design by Britnev to make his own icebreaker, *Eisbrecher I*.

The first true modern sea-going icebreaker was built at the turn of the 20th century. This vessel was the icebreaker *Yermak*, which was built in 1897 at the Armstrong Whitworth naval yard in England under contract from the Imperial Russian Navy. The ship borrowed the main principles from *Pilot* and applied them to the creation of the first polar icebreaker, which was able to run over and crush pack ice. The ship displaced 5,000 tonnes, and its steam-reciprocating engines delivered 10,000 horsepower (7,500 kW). The ship was decommissioned in 1963 and was scrapped in 1964, making it one of the longest-serving icebreakers in history. At the beginning of the 20th century, several other countries began to operate purpose-built icebreakers. Most were coastal icebreakers, but Canada, Russia, and later, the Soviet Union, also built several oceangoing icebreakers up to 11,000 tonnes in displacement. Before the first diesel-electric icebreakers were built in the 1930s, icebreakers were either coal- or oil-fired steamships. Reciprocating steam engines were preferred in icebreakers due to their reliability, robustness, good torque characteristics, and ability to reverse the direction of rotation quickly. During the steam era, the most powerful pre-war steam-powered icebreakers had a propulsion power of about 10,000 shaft horsepower (7,500 kW). The world's first diesel-electric icebreaker was the 4,330 tonne Swedish icebreaker *Ymer* in 1933. At 9,000 hp (6,700 kW) divided between two propellers in the stern and one propeller in the bow, she remained the most powerful Swedish icebreaker

Figure 21.1 USCG Polar Sea.

until the commissioning of *Oden* in 1957. *Ymer* was followed by the Finnish *Sisu*, the first diesel-electric icebreaker in Finland, in 1939. Both vessels were decommissioned in the 1970s and replaced by much larger icebreakers in both countries, the 1976-built *Sisu* in Finland and the 1977-built *Ymer* in Sweden (Figure 21.1).

In 1941, the US started building the Wind class icebreaker. Research in Scandinavia and the Soviet Union led to a design that had a very strongly built short and wide hull, with a cutaway forefoot and a rounded bottom. Powerful diesel-electric machinery drove two stern and one auxiliary bow propeller. These features would become the standard for post-war icebreakers until the 1980s. Since the mid-1970s, the most powerful diesel-electric icebreakers have been the formerly Soviet and later Russian icebreakers *Ermak*, *Admiral Makarov*, and *Krasin*, which each have nine 12-cylinder diesel generators producing electricity for three propulsion motors with a combined output of 26,500 kW (35,500 hp). In the 2020s, these will be surpassed by the new Canadian polar icebreaker, *CCGS John G. Diefenbaker*, which will have a combined propulsion power of 36,000 kW (48,000 hp). In Canada, diesel-electric icebreakers started to be built in 1952, starting with the *HMCS Labrador* (which was transferred later to the Canadian Coast Guard), using the USCG Wind-class design but without the bow propeller. Then in 1960, the next step in the Canadian development of large icebreakers came when *CCGS John A. Macdonald* was completed at Lauzon, Quebec. A bigger and more powerful ship than the *HMCS Labrador*, *CCGS John A. Macdonald* was an ocean-going icebreaker able to meet the most rigorous polar conditions. Her diesel-electric machinery of 15,000 horsepower (11,000 kW) was arranged in three units transmitting power equally to each of the three shafts. Canada's largest and most powerful icebreaker, the 120 m (390 ft) *CCGS Louis S. St-Laurent*, was delivered in 1969. Her original three steam turbine, nine generator, and three electric motor system produced some 27,000 shaft horsepower (20,000 kW). A multi-year mid-life refit project (1987–1993) saw the ship receive a new bow and a new propulsion system. The new power plant consists of five diesels, three generators, and three electric motors, providing the same propulsion power. On 22 August 1994, *Louis S. St-Laurent* and *USCGC Polar Sea* became the first North American surface vessels to reach the North Pole. The vessel was originally scheduled to be decommissioned in 2000; however, a refit extended the decommissioning date to 2017. It is now planned to be kept in service through the 2020s pending the introduction of a new class of polar icebreakers for the Coast Guard.

NUCLEAR ICEBREAKERS

Russia currently operates all existing and functioning nuclear-powered icebreakers. The first one, *NS Lenin*, was launched in 1957 and entered operation in 1959, before being officially decommissioned in 1989. It was both the world's first nuclear-powered surface ship and the first nuclear-powered civilian vessel. The second Soviet nuclear icebreaker was *NS Arktika*, the lead ship of the *Arktika class*. In service since 1975, she was the first surface ship to reach the North Pole, on 17 August 1977. Several nuclear-powered icebreakers were also built outside the Soviet Union. Two shallow-draft *Taymyr class* nuclear icebreakers were built in Finland for the Soviet Union in the late 1980s. In May 2007, sea trials were completed for the nuclear-powered Russian icebreaker *NS 50 Let Pobedy*. The vessel was put into service by the Murmansk Shipping Company, which manages all eight Russian state-owned nuclear icebreakers. The keel was originally laid in 1989 by Baltic Works of Leningrad, and the ship was launched in 1993 as *NS Ural*. This icebreaker was intended to be the sixth and last of the *Arktika class* and is currently the world's largest icebreaker (Figure 21.2).

Figure 21.2 Russian nuclear icebreaker *Arktika*.

FUNCTION

Today, most icebreakers are needed to keep trade routes open where there are either seasonal or permanent ice conditions. While the merchant vessels calling ports in these regions are strengthened for navigation on ice, they are usually not powerful enough to manage the ice by themselves. For this reason, in the Baltic Sea, the Great Lakes and the St. Lawrence Seaway, and along the Northern Sea Route, the main function of icebreakers is to escort convoys of one or more ships safely through ice-filled waters. When a ship becomes immobilised by ice, the icebreaker must free it by breaking the ice surrounding the ship and, if necessary, open a safe passage through the ice field. In difficult ice conditions, the icebreaker can also tow the weakest ships. Some icebreakers are also used to support scientific research in the Arctic and Antarctic. In addition to icebreaking capability, the ships need to have good open-water characteristics for transit to and from the polar regions, facilities and accommodation for the scientific personnel, and cargo capacity for supplying research stations on the shore. Countries such as Argentina and South Africa, which do not require icebreakers in domestic waters, have research icebreakers for conducting studies in the polar regions. As offshore drilling moves to the Arctic seas, icebreaking vessels are needed to supply cargo and equipment to the drilling sites and protect the drillships and oil platforms from ice by performing ice management, which includes, for example, breaking drifting ice into smaller floes and steering icebergs away from the protected asset. In the past, such operations were conducted primarily in North America, but today, Arctic offshore drilling and oil production is also going on in various parts of the Russian Arctic. The US Coast Guard uses icebreakers to help conduct search and rescue missions in the icy, polar oceans. US icebreakers serve to defend economic interests and maintain the presence of United States in the Arctic and Antarctic regions. As the icecaps in the Arctic continue to melt, there are more passageways being discovered. These navigation routes cause an increase of interests in the polar hemispheres from countries all over the world. The US polar icebreakers must continue to support scientific research in the expanding Arctic and Antarctic oceans. Every year, a heavy icebreaker must perform *Operation Deep Freeze*, clearing a safe path for resupply ships to the National Science Foundation's facility, McMurdo, in Antarctica. The most recent multi-month excursion was led by the *Polar Star*, which escorted a container and fuel ship through treacherous conditions before maintaining the channel free of ice. Without a heavy icebreaker, America would not be able to continue its polar research in Antarctica, as there would be no way to reach the science foundation.

CHARACTERISTICS

Icebreakers are often described as ships that drive their sloping bows onto the ice and break it under the weight of the ship. This only happens in very thick ice where the icebreaker will proceed at a walking pace or may even have to repeatedly back down several ship lengths and ram the ice pack at full power. More commonly the ice, which has a low flexural strength, is easily broken and submerged under the hull without a noticeable change in the icebreaker's trim while the vessel moves forward at a high and constant speed. When an icebreaker is designed, one of the main goals is to minimise the forces resulting from crushing, breaking the ice, and submerging the broken floes under the vessel. The average value of the longitudinal components of these instantaneous forces is called the ship's ice resistance. Naval architects who design icebreakers use the so-called *h-v-curve* to determine the icebreaking capability of the vessel. It shows the speed (v) that the ship can achieve as a function of ice thickness (h). This is done by calculating the velocity at which the thrust from the propellers equals the combined hydrodynamic and ice resistance of the vessel. An alternative means to determine the icebreaking capability of a vessel in different ice conditions such as pressure ridges is to perform model tests in an ice tank. Regardless of the method, the actual performance of new icebreakers is verified in full-scale ice trials once the ship has been built. To minimise the icebreaking forces, the hull lines of an icebreaker are usually designed so that the flare at the waterline is as small as possible. As a result, icebreaking ships are characterised by a sloping or rounded stem as well as sloping sides and a short parallel midship to improve manoeuverability in ice. However, the spoon-shaped bow and round hull have poor hydrodynamic efficiency and seakeeping characteristics and make the icebreaker susceptible to slamming, or the impacting of the bottom structure of the ship onto the sea surface. For this reason, the hull of an icebreaker is often a compromise between minimum ice resistance, manoeuverability in ice, low hydrodynamic resistance, and adequate open water characteristics.

Some icebreakers have a hull that is wider in the bow than in the stern. These so-called 'reamers' increase the width of the ice channel and thus reduce frictional resistance in the aft section as well as improve the ship's manoeuverability in ice. In addition to low friction paint, some icebreakers use an explosion-welded abrasion-resistant stainless steel ice belt that further reduces friction and protects the ship's hull from corrosion. Auxiliary systems such as powerful water deluges and air bubbling systems are used to reduce friction by forming a lubricating layer between the hull and the ice. Pumping water between tanks on both sides of the vessel results in continuous rolling that reduces friction and makes progress through the ice easier. Experimental bow designs such as the flat Thyssen-Waas bow and a cylindrical bow have been tried over the years to further reduce the ice resistance and create an ice-free channel.

STRUCTURAL DESIGN

Icebreakers and other ships operating in ice-filled waters require additional structural strengthening against various loads resulting from the contact between the hull of the vessel and the surrounding ice. As ice pressures vary between different regions of the hull, the most reinforced areas in the hull of an ice-going vessel are the bow, which experiences the highest ice loads, and around the waterline, with additional strengthening both above and below the waterline to form a continuous ice belt around the ship. Short and stubby icebreakers are built using transverse framing in which the shell plating is stiffened with frames placed about 400–1,000 mm (1–3 ft) apart as opposed to longitudinal framing used in longer ships. Near the waterline, the frames running in a vertical direction distribute the locally concentrated ice loads on the shell plating to longitudinal girders called stringers, which in turn are supported by web frames and bulkheads that carry the more spread-out hull loads. While the shell plating, which is in direct contact with the ice, can be up to 50 mm (2 in) thick in older polar icebreakers, the use of high-strength steel with yield strength up to 500 MPa (73,000 psi) in modern icebreakers results in the same structural strength with smaller material thicknesses and lower steel weight. Regardless of the strength, the steel used in the hull structures of an icebreaker must be capable of resisting brittle fracture in low ambient temperatures and high loading conditions, both of which are typical for operations in ice-filled waters. If built according to the rules set by a classification society such as the American Bureau of Shipping, Det Norske Veritas or Lloyd's Register, icebreakers may be assigned an ice class based on the level of ice strengthening in the ship's hull. It is usually determined by the maximum ice thickness where the ship is expected to operate and other requirements

Figure 21.3 Russian icebreaker *Tor*.

such as limitations on ramming. While the ice class is generally an indication of the level of ice strengthening, not the actual icebreaking capability of an icebreaker, some classification societies such as the Russian Maritime Register of Shipping have operational capability requirements for certain ice classes. Since the 2000s, the International Association of Classification Societies (IACS) has proposed adopting a unified system known as the Polar Class (PC) to replace classification society-specific ice class notations (Figure 21.3).

POWER AND PROPULSION

Since World War II, most icebreakers have been built with diesel-electric propulsion in which diesel engines coupled to generators produce electricity for propulsion motors that turn the fixed pitch propellers. The first diesel-electric icebreakers were built with direct current (DC) generators and propulsion motors, but over the years, the technology advanced first to alternating current (AC) generators and finally to frequency-controlled AC-AC systems. In modern diesel-electric icebreakers, the propulsion system is built according to the power plant principle in which the main generators supply electricity for all onboard consumers and no auxiliary engines are needed. Although the diesel-electric powertrain is the preferred choice for icebreakers due to the good low-speed torque characteristics of the electric propulsion motors, icebreakers have also been built with diesel engines mechanically coupled to reduction gearboxes and controllable pitch propellers. The mechanical powertrain has several advantages over diesel-electric propulsion systems, such as lower weight and better fuel efficiency. However, diesel engines are sensitive to sudden changes in propeller revolutions, and to counter this, mechanical powertrains are usually fitted with large flywheels or hydrodynamic couplings to absorb the torque variations resulting from propeller-ice interaction. The 1969-built Canadian polar icebreaker *CCGS Louis S. St-Laurent* was one of the few icebreakers fitted with steam boilers and turbogenerators that produced power for three electric propulsion motors. It was later refitted with five diesel engines, which provide better fuel economy than steam turbines. Later, Canadian icebreakers were built with diesel-electric powertrains. Two PC icebreakers operated by the US Coast Guard have a combined diesel-electric and mechanical propulsion system that consists of six diesel engines and three gas turbines. While the diesel engines are coupled to generators that produce power for three propulsion motors, the gas turbines are directly coupled to the propeller shafts driving controllable pitch propellers. The diesel-electric power plant can produce up to 13,000 kW (18,000 hp), while the gas turbines have a continuous combined rating of 45,000 kW (60,000 hp).

The number, type, and location of the propellers depend on the power, draft, and intended purpose of the vessel. Smaller icebreakers and icebreaking special purpose ships may be able to do with just one propeller while large polar icebreakers typically need up to three large propellers to absorb all power and deliver enough thrust. Some shallow draught river icebreakers have been built with four propellers in the stern. Nozzles may be used to increase the thrust at lower speeds, but they may become clogged by ice. Until the 1980s, icebreakers operating regularly in ridged ice fields in the Baltic Sea were fitted with the first one and later two bow propellers to create a powerful flush along the hull of the vessel. This increased the icebreaking capability of the vessels by reducing the friction between the hull and the ice and allowed the icebreakers to penetrate thick ice ridges without ramming. However, the bow propellers are not suitable for polar icebreakers operating in the presence of harder multi-year ice and thus have not been used in the Arctic. Azimuth thrusters remove the need of traditional propellers and rudders by having the propellers in steerable gondolas that can rotate 360 degrees around a vertical axis. These thrusters improve propulsion efficiency, icebreaking capability, and manoeuverability of the vessel. The use of azimuth thrusters also allows a ship to move astern in ice without losing manoeuverability. This has led to the development of double-acting ships, vessels with the stern shaped like an icebreaker's bow, and the bow designed for open water performance. In this way, the ship remains economical to operate in open water without compromising its ability to operate in difficult ice conditions. Azimuth thrusters have also made it possible to develop new experimental icebreakers that operate sideways to open a wide channel through the ice.

POLAR CLASS ICEBREAKERS

As mentioned above, PC refers to the ice class assigned to a ship by a classification society based on the Unified Requirements for PC Ships developed by the IACS. Seven PCs are defined in the rules, ranging from PC1 for year-round operation in all polar waters to PC7 for summer and autumn operation in thin first-year ice. The IACS PC rules should not be confused with the *International Code for Ships Operating in Polar Waters* (Polar Code) mandated by the IMO. The development of the PC rules began in the 1990s with an international effort to harmonise the requirements for marine operations in the polar waters to protect life, property, and the environment. The guidelines developed by the IMO, which were later incorporated in the Polar Code, referred to the compliance with Unified Requirements for Polar Ships developed by the IACS. In May 1996, an 'Ad-Hoc Group to establish *Unified Requirements for Polar Ships* (AHG/PSR)' was established with one working group concentrating on the structural requirements and another working on machinery-related issues. The first IACS PC rules were published in 2007. Prior to the development of the unified requirements, each classification society had its own set of ice class rules ranging from Baltic ice classes intended for operation in first-year ice to higher vessel categories, including icebreakers, intended for operations in polar waters. When developing the upper and lower boundaries for the PCs, it was agreed that the highest *Polar Class* vessels (PC1) should be capable of operating safely anywhere in the Arctic or the Antarctic waters at any time of the year, while the lower boundary was set to existing tonnage operating during the summer season, most of which followed the Baltic ice classes with some upgrades and additions. The lowest PC (PC7) was thus set to a similar level to the Finnish-Swedish ice class 1A. The definition of operational conditions for each PC was intentionally left vague due to the wide variety of ship operations conducted in polar waters (Figure 21.4).

Polar class notations

IACS has established seven different PC notations, ranging from PC1 (the highest) to PC7 (the lowest), with each level corresponding to the operational capability and strength of the vessel. The descriptions given in the rules are intended to guide owners, designers, and administrations in selecting the appropriate PC to match the intended voyage or service of the vessel. Ships with sufficient power and strength to undertake 'aggressive operations in ice-covered waters', such as escort and ice management operations, can be assigned an additional notation 'Icebreaker'. The lowest PCs (PC6 and PC7) are equivalent to the two highest Finnish-Swedish ice classes (1A Super and 1A, respectively). However, unlike the Baltic ice classes intended for operation only in first-year sea ice, even the lowest PCs consider the possibility of encountering multi-year ice (') (Table 21.1).

Figure 21.4 Japanese icebreaker *Shirase*.

Table 21.1 IACS polar class notations

Polar class	*Ice description (based on WMO sea ice nomenclature)*
PC 1	Year-round operation in all polar waters
PC 2	Year-round operation in moderate multi-year ice conditions
PC 3	Year-round operation in second-year ice, which may include multi-year ice inclusions
PC 4	Year-round operation in thick first-year ice, which may include old ice inclusions
PC 5	Year-round operation in medium first-year ice, which may include old ice inclusions
PC 6	Summer/autumn operation in medium first-year ice, which may include old ice inclusions
PC 7	Summer/autumn operation in thin first-year ice, which may include old ice inclusions

Requirements

In the PC rules, the hull of the vessel is divided longitudinally into four regions: the bow, the bow intermediate, the midbody, and the stern. All longitudinal regions except the bow are further divided vertically into the bottom, the lower, and the ice belt regions. For each region, a design ice load is calculated based on the dimensions, hull geometry, and ice class of the vessel. This ice load is then used to determine the scantlings and steel grades of structural elements such as shell plating and frames in each location.

The design scenario used to determine the ice loads is a glancing impact with ice. In addition to structural details, the PC rules have requirements for machinery systems such as the main propulsion, steering gear, and systems essential for the safety of the crew and survivability of the vessel. For example, propeller-ice interaction should be considered in the propeller design, cooling systems and seawater inlets should be designed to work also in ice-covered waters, and the ballast tanks should be provided with effective means of preventing freezing. Although the rules require the ships to have suitable hull form and sufficient propulsion power to operate independently and at continuous speed in ice conditions corresponding to their PC, the ice-going capability requirements of the vessel are not clearly defined in terms of speed or ice thickness. In practice, this means that the PC of the vessel does reflect the actual icebreaking capability of the vessel.

POLAR CLASS SHIPS

The IACS PC rules apply for ships contracted for construction on or after 1 July 2007. This means that while the vessels built prior to this date may have an equivalent or even a higher level of ice strengthening, they are not officially assigned a PC and may not in fact fulfil all the requirements in the unified requirements. In addition, particularly Russian ships and icebreakers are assigned ice classes only according to the requirements of the Russian Maritime Register of Shipping, which maintains its own ice class rules parallel to the IACS PC rules.

Polar Class 5 and below

Although several ships have been built to the two lowest PCs (such as the Peruvian PC7 *Carrasco*), corresponding to the two highest Baltic ice classes, there are only a handful of vessels assigned ice class PC5 or higher. While the 2012-built drillship *Stena IceMAX* has a hull strengthened according to PC4 requirements, the vessel is not capable of independent operation in ice and has ice class PC6. The South African polar research vessel *S. A. Agulhas II*, also delivered in 2012, was the first ice-capable vessel built to ice class PC5. In 2012, the Royal Canadian Navy awarded a contract to construct several offshore patrol vessels, the *Harry DeWolf class* offshore patrol vessel, with a hull designed to meet PC5+ requirements. Eight ships were planned to be built, six for the navy, and two for the Canadian Coast Guard. The first ship was commissioned in 2021.

Polar Class 4

The Canadian-flagged icebreaking bulk carrier *Nunavik*, operated by Fednav and used to transport copper and nickel from the Nunavik Nickel

Project, was built to ice class PC4 in 2014. The Finnish icebreaker *Polaris* was built in 2016 to the same ice-class with additional Lloyd's Register class notation 'Icebreaker (+)' where the last part refers to additional structural strengthening based on analysis of the vessel's operational profile and potential ice loading scenarios. *RRS Sir David Attenborough*, operated by British Antarctic Survey, has a PC4 hull and a PC5 propulsion system.

Polar Class 3

The Netherlands-based ZPMC-Red Box Energy Services operates two PC3 class deck cargo ships, *Audax* and *Pugnax*, both of which were built in 2016. The Norwegian icebreaking polar research vessel, *Kronprins Haakon* and the Chinese Research Vessel, *Xue Long 2* are both rated PC3. The Australian Icebreaker, *RSV Nuyina*, is rated PC3 Icebreaker (+).

Polar Class 2 and above

In December 2017, the French cruise ship operator Compagnie du Ponant announced an order for an icebreaking PC2 ice class cruise ship capable of taking tourists to the North Pole. The vessel was delivered in July 2021. By late 2021, the US Coast Guard had ordered two out of three planned heavy polar icebreakers referred to as Polar Security Cutters with a PC2 rating. The proposed Canadian polar icebreaker for the Canadian Coast Guard, *CCGS John G. Diefenbaker*, is designed to ice class PC2 Icebreaker (+) class standards.

As of 2022, no ships have been built to PC1, the highest ice class specified by IACS.

Chapter 22

Offshore support vessels

Offshore vessels (OSVs) are ships that specifically serve operational purposes such as oil exploration and construction work out at sea. There are a variety of OSVs, which not only help in the exploration and drilling of oil and gas but also for providing necessary supplies for offshore installations such as food and fresh water, machinery and replacements, chemicals, mud, and a variety of other products that are needed to keep the offshore oil and gas industry operating. Offshore ships also provide services delivering and relieving personnel to and from the platforms. As mentioned above, OSVs are a collective of many diverse types of vessels, though they can be classified into the following main categories:

- Oil exploration and drilling vessels;
- Offshore support vessels;
- Offshore production vessels; and
- Construction/Special purpose vessels.

Each of these categories comprises a variety of vessels. For example, oil exploration vessels, as the name suggests, help in the exploration and drilling of offshore oil and gas. The main types of exploration vessels are drill ships, jack-up vessels, semi-submersible vessels, offshore barges, floating platforms, and tenders. Certain OSVs provide the necessary manpower and technical reinforcement required so that the operational processes of these platforms continue smoothly and without interruption. These vessels are referred to as offshore support vessels or offshore supply vessels. Offshore supply vessels transport the required structural components to the designated offshore sector along as well as supply freight and personnel. The constructional aspect of these vessels is purpose-built to suit specific operational demands, for example, having a helicopter landing deck or a fully manoeuverable gear (crane). Some of the main types of offshore support vessels are as follows:

- Anchor Handling Tug Supply (AHTS) vessels;
- Seismic vessel;

DOI: 10.1201/9781003342366-27

- Platform supply vessels (PSVs);
- Well intervention vessels;
- Accommodation ships; and
- Offshore production vessels.

Offshore production vessels refer to those vessels that help in the production process conducted by drilling units. FPSOs, which we covered earlier, are an excellent example of ships that can be used in the production of offshore oil and gas. In addition to FPSOs, some of the other types of vessels included in this category are single point anchor reservoir (SPAR) platforms, shuttle tankers, and tension leg platforms (TLPs). Ships that primarily aid in the construction of various high seas structures are referred to as offshore construction vessels (OCVs). Other OSVs of these type also include those that provide anchorage and tugging assistance and those types of ships that help in the positioning of deep sub-water cable and piping lines. These vessels include diving support vessels (DSVs), crane vessels (CVs), and pipe laying vessels (PLVs). In addition to these, those ships that provide aid in case of emergencies at sea and those types of vessels that undertake research and analysis activities are also included in the OSV category. The ever-growing need to explore and suitably harness the valuable commodities beneath the seabed has led to substantial growth in the need and demand for offshore ships. Coupled with the advantages of scientific research and technological advancement in offshore maritime technology, the present-day fleet of OSVs is a clear portrayal of the huge strides taken by the maritime sector.

PLATFORM SUPPORT VESSELS

Platform supply vessels (PSVs) are a type of ship specially designed to supply offshore oil and gas platforms. These ships range from 50 to 100 m (160–330 ft) in length and accomplish a variety of tasks. The primary function of most of these vessels is logistic support and transportation of goods, tools, equipment, and personnel to and from offshore oil platforms and other offshore structures. In recent years, a new generation of PSVs entered the market, usually equipped with Class 1 or Class 2 dynamic positioning systems. Military applications are also under development as with the Status-6 Oceanic Multipurpose System. PSVs belong to the broad category of OSVs that include PSVs, CVs and well stimulation vessels (WSVs), anchor handling tug supply vessels (AHTS), and OCVs. Larger OSVs have extensive sophisticated equipment, including remotely operated underwater vehicles (ROVs), and tend to accommodate a larger number of personnel, often in excess of 100 people. A primary function of the PSV is to transport supplies to the oil platform and return other cargoes to the shore. Cargo tanks for drilling mud, pulverised cement, diesel fuel, potable and non-potable water, and chemicals used in the drilling process comprise the bulk of the cargo spaces. Fuel, water, and chemicals are always required by oil platforms. Certain other chemicals must be returned to shore for proper recycling or disposal; however, crude oil product from the rig is usually not a supply vessel cargo. Common and specialty tools are transported on the large decks of these vessels. Most carry a combination of deck cargoes and bulk cargo in tanks below the deck. Many ships are constructed (or re-fitted) to accomplish a particular job. Some of these vessels are equipped with a firefighting capability and fire monitors for fighting platform fires. Some vessels are equipped with oil containment and recovery equipment to assist in the clean-up of a spill at sea. Other vessels are equipped with tools, chemicals, and personnel to 'work over' existing oil wells for the purpose of increasing the wells' production (Figure 22.1).

The crew on these ships can number up to 36 members, depending on the size, working area, and whether the vessel is DP equipped or not. CVs and drillships often have 100–200 people on board including a resolute project team. Crews sign on to work and live aboard the ship for an extended period, which is often followed by a similar period of shore leave. Depending on the ship's owner or operator, the time on board varies from 1 to 3 months with 1 month off. Work details on PSVs, like many ships, are organised into shifts of up to 12 hours. Living on board the ship, each crew member and worker will have at least a 12-hour shift, lasting some portion of a 24-hour day. Supply vessels are provided with a 'bridge' area for navigating and operating the ship, machinery spaces, living quarters, galley, and mess room.

Figure 22.1 Offshore support vessel *Balder Viking*, arriving Aberdeen Harbour, Scotland.

Some have built-in work areas and communal areas for entertainment. The large main deck area may sometimes be used for portable housing. Living quarters consist of cabins, lockers, offices, and spaces for storing personal items. Living areas are provided with wash basins, showers, and toilets. The galley or cooking and eating areas on board ship will not only be stocked with enough supplies to last the intended voyage but also with the ability to store provisions for several more months if required. A walk-in size cooler and freezer, a commercial stove and oven, deep sinks, storage, and counter space are available for the crew member responsible for food preparation and cooking. Eating areas typically have coffee machines, toasters, microwave ovens, cafeteria-style seating, and other amenities needed to feed a diligent crew in often difficult conditions.

ANCHOR HANDLING TUG SUPPLY VESSELS

Anchor handling tug supply (AHTS) vessels are built to manage anchors for oil rigs, tow them to locations, and use them to secure the rigs in place. AHTS vessels sometimes also serve as emergency response and rescue vessels (ERRVs) and as supply transports. Many of these vessels are designed to meet the harsh conditions of the North Sea and can undertake supply duties between land bases and drilling sites. They also provide towing assistance during tanker loading, deep water anchor handling, and towing of threatening objects. AHTS vessels differ from PSVs in being fitted with winches for towing and anchor handling, having an open stern to allow the decking of anchors, and having more power to increase the bollard pull. The machinery is specifically designed for anchor handling operations. They also have arrangements for quick anchor release, which is operable from the bridge or other normally manned location in direct communication with the bridge. The reference load used in the design and testing of the towing winch is twice the static bollard pull. Even if AHTS vessels are customised for anchor-handling and towing, they can also undertake, for example, ROV (remotely operated underwater vehicle) services, safety and rescue services, and supply duties between mainland and offshore installations. During anchor handling operations, the heavy machinery and related equipment are repeatedly pushed to their extreme limits in very hostile environments. This means breakdowns do happen, although not frequently. Because of rough seas, it is common for the vessel to inadvertently brush against the installation, or to push up against the high-tension wires and chains. AHTS are purpose-built with strengthened hulls and are completely different from typical harbour tugs, and even ocean-going tugs. The latest anchor handling tugs are designed to withstand the toughest maritime conditions, keeping in mind safety, comfort, and crew effectiveness. This makes them the superior choice for operations in remote offshore oil and gas fields. All AHTS vessels have certain attributes and characteristics in common, including:

- A superior bollard pull and a higher engine rating (BHP), which makes them powerful enough for specialised jobs such as anchor handling;
- A combination of multiple thrusters (bow and stern) with twin-screw CPP systems, providing excellent vessel handling features that allow such vessels to work in almost any sea condition;
- A large volume of strengthened deck space astern of the accommodation areas allows even the largest of the anchors, heavy wires, chains, buoys, and other related equipment to be stowed and overseen.
- An extremely powerful multi-drum system catering twin winches, one each for towing and anchor handling purposes. These are separate from the combination of other spare drums and work winches used especially for towing and deep-water anchor handling.
- Sufficient anchor chains can be stowed on board these vessels due to the availability of larger capacity chain lockers.

Despite their name, AHTS are multipurpose vessels that can even perform the duties of ordinary PSVs, such as carrying large quantity of water, fuel, and deck cargo. Similarly, the AHTS vessels can be used in many of the diverse applications required of support ships servicing the offshore oil and gas industry. As well as performing towing operations, rig moves, and anchor work for rigs, semi-submersibles, and construction barges, AHTS vessels execute general supply duties by carrying dry and liquid cargo such as cement, mud, fresh water, and fuel oil. If ocean-going tugs are not readily available, then the AHTS vessels can be used instead for such duties as salvaging and rescue. Today, AHTS are progressively being used for towing and anchor handling of newer offshore structures such as the tension leg platforms (TLPs) and gravity-based platforms (Figure 22.2).

Figure 22.2 Anchor handling tugs at Scrabster, Scotland.

DIVING SUPPORT VESSELS

A diving support vessel is a type of offshore ship that is used as a floating base for professional diving projects. Commercial diving support vessels emerged during the 1960s and 1970s, when the need arose for offshore diving operations to be performed below and around oil production platforms and associated installations in open water in the North Sea and the Gulf of Mexico. Until that point, most diving operations were carried out from mobile oil drilling platforms, pipe-lay, or crane barges. The diving system tended to be modularised and craned on and off the vessel as a package. As permanent oil and gas production platforms emerged, the owners and operators were not keen to give over valuable deck space to diving systems because after they came on-line, the expectation of continuing diving operations was low. However, as equipment fails or gets damaged, and there was a regular, if not continuous, need for diving operations in and around oil fields, it was quickly realised that some form of offshore diving capability was needed. The solution was to put diving packages on ships. Initially, these tended to be oilfield supply ships or fishing vessels; however, keeping these types of ships 'on station,', particularly during inclement weather, made the diving operation dangerous, problematic, and seasonal. Furthermore, seabed operations usually entailed the raising and lowering of heavy equipment, and most such vessels were not equipped for this type of operation. This is when the dedicated commercial diving support vessel started to emerge. These were often built from scratch or heavily converted pipe carriers or other utility-type ships (Figure 22.3).

The key components of the diving support vessel are (1) dynamic positioning or DP. Controlled by a computer with input from position reference systems (DGPS, transponders, light taut wires, or RadaScan), the vessel will maintain the ship's position over a dive site by using multi-directional thrusters, whilst other sensors compensate for swell, tide, and prevailing winds; and (2) saturation diving systems. For diving operations below 50 m (164 ft), a mixture of helium and oxygen (heliox) is required to eliminate the narcotic effect of nitrogen when breathed in under pressure. For extended diving operations at depth, saturation diving is the preferred approach. A saturation system is installed on the ship. A diving bell transports the divers between the saturation system and the work site, which is lowered through a 'moon pool' at the bottom of the ship, usually with a support structure 'cursor' to support the diving bell through the turbulent waters near the surface. There are several support systems for the saturation system on a diving support vessel, usually including a remotely operated vehicle (ROV) and heavy lifting equipment. Most of the vessels currently operating in the North Sea were built in the 1980s. The semi-submersible fleet, the Uncle John, and similar have proven to be too expensive to maintain and too slow to move between fields. Therefore, most existing designs are monohull vessels with either a one or a twin bell dive system. Although there was little innovation between the 1980s

Figure 22.3 Diving support vessel *Kingfisher*.

and early 2000s, high oil prices around 2004 spurred the market for subsea developments in the North Sea significantly. This led to a scarcity of diving support vessels. In response, contractors ordered several new build vessels, which entered the market in 2008. These vessels are built and designed not only to support diving activities, but also to support ROV operations with dedicated hangar and LARS for ROVs, seismic survey operations, and cable-laying operations, etc. Owing to these modern-day vessels, they may have at any time 80–150 project personnel on board, including divers, diving supervisors and superintendents, dive technicians, life support technicians and supervisors, ROV pilots, ROV superintendents, survey teams, and of course the client's representatives. For all these personnel to conduct their contracted job with an oil and gas company, a professional crew navigate and operate the vessel as per the requirements and instructions of the diving or ROV or survey team superintendents. However, the ultimate responsibility lies with the master of the vessel for the safety of everyone on board. In expanding the utility of the vessel, just like liveaboard dive boats, these vessels, in addition to the usual domestic facilities expected by hotel guests, have specialised mix gas diving compressors and reclaim systems, gas storage and gas blending facilities, as well as purpose-built saturation chambers where the divers in compression live. These vessels are designed to be hired by diving service-providing companies or directly by oil and gas contractors who then hire a diving or ROV or survey service-providing company. This company uses the vessel as a platform for performing their contracted duties. Despite diving from a DSV makes it possible to undertake a wider range of operations, the platform presents some inherent hazards, and equipment and procedures must be adopted to manage these hazards as well as the hazards of the environment and diving tasks.

EMERGENCY TOW VESSELS

An emergency tow vessel, also known as emergency towing vessel (ETV), is a multipurpose ship used by state authorities to tow disabled vessels on the high seas to prevent further hazards to shipping and the marine environment. The disabled vessel is either towed to a haven or kept in place against wind and current until commercial assistance by tugboats can arrive on site or until the vessel has been repaired to the extent of being able to manoeuvre on its own. The need for ETVs as a preventive measure has arisen since the number of available commercial salvage tugs was reduced while potential dangers from individual vessels have increased. Globally, there are more ETVs positioned in and around the European continent coastline than anywhere else in the world. For example, Spain has 14 ETVs, Turkey has 11, Germany operates eight, Norway has seven, France has five, Sweden has three, and the Netherlands, Poland, South Africa, Iceland, and Finland each have one official emergency tugboat. Australia also operates emergency response vessels. The UK's four-strong ETV fleet was to be disbanded in September 2011 due to budget cuts, but the two vessels operating in Scottish waters received an extension of the contract until the end of 2011. As of 2022, the UK has no government-owned ETV capability. Distressed vessels that request the service of towing vessels have the means to make towing as safe as possible. Oil tankers have emergency towing equipment fixed at the forward and aft part of the vessel that will allow connecting the towing line. The connection of these apparatuses to the vessel's hull is reinforced according to class requirements. Bulk carriers and general cargo vessels are not required to have a specialised emergency towing arrangement. Depending on the vessel's type and keel laid date, in accordance with the MSC 256(84) standard, they must have on board an emergency towing procedure manual. This ship-specific manual describes procedures that will allow the vessel to be towed using its own equipment. The procedure should make use of the standard mooring equipment like mooring ropes, bits, rollers, and Panama chock (Figure 22.4).

National ETV fleets

Algeria

In 2010, Algeria ordered three tugboats of the *Bourbon class*, which were already in use by the French harbours at Brest and Cherbourg. The ships

Figure 22.4 ERRV *Grampian Deliverance*, departing Aberdeen Harbour, Scotland.

were built by STX OSV in Norway and STX Tulcea in Romania. The first vessel *El Moundjid* was ready for delivery in December 2011, with the two others scheduled for delivery in June and September 2012. With a bollard pull of 200 metric tonnes and a speed of 20 kts (23 mph; 37 km/h), the Algerian ETVs are an improved version of the French *Bourbon class*. They are based in Oran and Skikda. By acquiring these three ships, Algeria became the leading Mediterranean nation in terms of marine salvage as of 2012.

Finland

The Ministry of the Environment operates the *YAG Louhi*. The ship is listed as a multipurpose oil recovery vessel and can be used for emergency towing, firefighting, icebreaking, mine-laying, oil and chemical spill response, as well as other rescue operations. The vessel has a bollard pull of 60 metric tonnes. *YAG Louhi* is based at the Port of Upinniemi approximately 24 mi (40 km) west of Helsinki in the Archipelago Sea.

France

For assistance and salvage, five ocean-going tugs and their crews are ready to respond around-the-clock. These are the *Abeille Bourbon*, based in Brest, with a bollard pull of 200 metric tonnes; the *Abeille Liberté*, based in Cherbourg, with a bollard pull of 200 metric tonnes; the *Abeille Flandre*, based in Toulon, with a bollard pull of 160 metric tonnes; the *Abeille Languedoc*, based in La Rochelle, with a bollard pull of 160 metric tonnes; and the *Jason*, also based in Toulon, with a bollard pull of 124.2 metric tonnes. The tugs are chartered by the French government and manned by a civilian crew.

Germany

Responsible for the German ETV flotilla is the Central Command for Maritime Emergencies (CCME), based in Cuxhaven. The German concept of emergency towing prescribes a maximum response time of two hours for any incident in German coastal waters. This requires three ETVs in the North Sea and five in the Baltic Sea despite having a smaller area to cover. The equipment and performance of the vessels have been adapted to the size of the vessels in the respective areas of operation and include the ability to operate in shallow waters. Moreover, it is mandatory to have one vessel with 200 metric tonnes of bollard pull and 100 metric tonnes each in the North Sea and Baltic, respectively. Both ship types are also required to be able to operate under hazardous conditions such as explosive areas and gas leaks. Four out of the eight German ETVs are multipurpose vessels owned by the Federal Waterways and Shipping Administration and are part of the German Federal Coast Guard, while another four have been chartered from tug companies. Since 2001 a cooperation agreement with the Netherlands comprised the Dutch ETV and the German *Nordic*, which replaced the *ETV Oceanic*. Operating in the North Sea are the *Nordic* (East Frisian Islands, based in Cuxhaven, bollard pull of 201 metric tonnes); *Mellum* [5 nmi (28 mi; 9.3 km)] southwest Heligoland, bollard pull of 100 metric tonnes); and the *Neuwerk*: 5 nmi (5 mi; 9.3 km) southwest Süderoogsand (Nordfriesland), bollard pull of 113 metric tonnes). Operating in the Baltic Sea are the vessels *Bülk* (Kiel Fjord, bollard pull of 40 metric tonnes); *Scharhörn* (Hohwacht Bay, between Kiel and Fehmarn, bollard pull of 40 metric tonnes); *Baltic* (Warnemünde, bollard pull of 127 metric tonnes); *Arkona* (Stralsund, bollard pull of 40 metric tonnes); and *Fairplay 25* (Sassnitz, Rügen, bollard pull of 65 metric tonnes).

Iceland

Iceland operates the *ICGV Þór* (English: *ICGV Thor*). With a bollard pull of approximately 110 metric tonnes, the Icelandic Coast Guard vessel is capable of towing-stricken tankers of up to about 200,000 metric tonnes dwt.

Netherlands

The Netherlands Coastguard operates one ETV on charter from Svitzer Wijsmuller, the *Ievoli Amaranth*, and is based in Den Helder.

Norway

The Norwegian Coast Guard owns three *Barentshav class* OPV multipurpose vessels and operates another multipurpose vessel named *NoCGV Harstad* on charter. Their main purpose is the prevention of pollution by oil tankers along the Norwegian coastline. Therefore, the ships can also be used in the ETV role with a bollard pull exceeding 100 metric tonnes. In addition, Norway also charters the *Beta*, bollard pull of 118 metric tonnes; the *Normand Jarl*, bollard pull of 150 metric tonnes; and the *North Crusader*, bollard pull of 144 metric tonnes.

Poland

The Polish Ministry of Transport operates one ETV on charter, the tug *Kapitan Poinc*, with a bollard pull of 74 metric tonnes.

South Africa

The tug *Smit Amandla* (bollard pull of 181 metric tonnes) is based in the port of Cape Town.

Spain

The Sociedad de Salvamento y Seguridad Marítima has a total of 14 multipurpose vessels for search and rescue (SAR) and pollution prevention duties. These are the *Don Inda Class*, consisting of two sister ETVs based on Ulstein's UT 722 L design (bollard pull of 228 metric tonnes); the *Luz de Mar Class*, comprising two sister ETVs (bollard pull of 128 metric tonnes); the *Alonso de Chaves*_ (bollard pull of 105 metric tonnes); the *Punta Salinas* (bollard pull of 97 metric tonnes); the *Punta Mayor* (bollard pull of 81 metric tonnes); and the *María de Maeztu Class*, comprising seven sister ETVs, each with a bollard pull of 60 metric tonnes.

Sweden

The Swedish Coast Guard operates three EVTs of the same type. Built by Damen, these multifunctional patrol and emergency response vessels have a bollard pull of 100 metric tonnes each. They were brought into operation in 2009 and 2010. The fleet consists of the *Poseidon* (KBV 001), based in Gothenburg; the *Triton* (KBV 002), based in Slite on Gotland; and *Amfitrite* (KBV 009), based in Karlskrona.

Turkey

The Turkish Directorate General of Coastal Safety operates 11 ETVs along with numerous SAR, oil spill response, and firefighting vessels throughout the Bosphorus and Dardanelles, where the organisation has a de jure monopoly for marine salvage along with the Sea of Marmara. Some of these ETVs also serve as escort tugs for vessels passing through Bosphorus and Dardanelles, which make up the Turkish Straits System, one of the busiest and most dangerous seaways in the world. The organisation also deals with navigational aids around the Turkish coastline, SAR operations, pilotage at both straits and some Turkish ports and, most importantly, the Turkish Straits Vessel Traffic Systems (VTS). The Turkish ETV fleet currently comprises the *Gemi Kurtaran*, bollard pull of 75 metric tonnes; the *Kurtarma 1*, bollard pull of 53 metric tonnes; *Kurtarma 2*, bollard pull of 53 metric tonnes; *Kurtarma 3*, bollard pull of 70 metric tonnes; *Kurtarma 4*, bollard pull of 70 metric tonnes; *Kurtarma 5*, bollard pull of 65 metric tonnes; *Kurtarma 6*, bollard pull of 66 metric tonnes; *Kurtarma 7*, bollard pull of 60 metric tonnes; *Kurtarma 8*, bollard pull of 60 metric tonnes; *Kurtarma 9*, bollard pull of 105 metric tonnes; *Kurtarma 10*, bollard pull of 105 metric tonnes; *Seyit Onbasi*, Oil spill response vessel; and the *Nene Hatun*, bollard pull of 205 metric tonnes.

United Kingdom

The UK's ETV vessels were chartered by the Maritime and Coastguard Agency (MCA) for use in pollution control or towing vessels that were in difficulty. The vessels were a combination of tugboat, anchor handler, fireboat, and buoy tender. As of 2010, four ETVs, *Anglian Prince*, *Anglian Princess*, *Anglian Sovereign*, and *Anglian Monarch*, were based in strategic locations around the United Kingdom, with two covering the south coast of England, at Falmouth in Cornwall and Dover in Kent, and two in Scottish waters, at Stornoway the Western Isles (the Outer Hebrides), and Lerwick in the Northern Isles (Shetland and Orkney). The four-strong ETV fleet was intended to be operational 24 hours a day, 365 days a year, and maintained at 30 minutes readiness to sail, with one tug allocated to each of the four operating areas on a rotational basis, accounting for maintenance schedules. The Dover station was funded jointly by French maritime authorities. A fifth tug, the *Anglian Earl*, was an anchor handling and salvage tug not only extensively used on commercial work, but also fitted the ETV criteria, and functioned as cover for any of the four ETV stations as and when required. In 2010, the Government announced, as part of the Department for Transport's share of cuts in the Comprehensive Spending Review, that the ETV fleet would no longer be funded by the MCA from September 2011, saving £32.5m over the Spending Review period. The Department stated that 'state provision of ETVs does not represent a correct use of taxpayers money and that ship salvage should be a commercial matter between a ship's operator and the salvor'. On 30 September 2011, it was announced that the two ETVs operating in the Minch and the Shetland Islands would receive a moratorium of three months with interim funding provided by the British government.

FIREBOATS

A fireboat or fire-float is a specialised type of watercraft with pumps and nozzles designed for fighting shoreline and shipboard fires. The first fireboats, dating to the late 18th century, were tugboats, retrofitted with firefighting equipment. Older designs derived from tugboats and modern fireboats more closely resembling seafaring ships can both be found in service today. These ships are frequently used for fighting fires on docks and shoreside warehouses, as they can directly attack fires in the supporting underpinnings of these structures. They also have an effectively unlimited supply of water available, pumping directly from below the hull. Fireboats can be used to assist shore-based firefighters when other water is in low supply or is unavailable, for example during the Loma Prieta earthquake in San Francisco, in 1989. Some modern fireboats are capable of pumping tens of thousands of gallons of water per minute. An example is the fireboat operated by the Los Angeles Fire Department, the *Warner Lawrence*, which has the capability to pump up to 38,000 US gallons per minute (2.4 m^3/s; 32,000 imp gal/min) and up to 122 m (400 ft) into the air. Fireboats are most usually seen by the public when welcoming a new cruise ship with a display of their water-moving capabilities, throwing large arcs of water in every direction. Occasionally, fireboats are used to carry firefighters, Emergency Medical Technicians, and medical doctors with their equipment to islands and other boats. In some regions, fireboats may be used as icebreakers, such as the Chicago Fire Department's *Victor L. Schlaeger*, which can break 20.3–30.4 cm (8–12 in) of ice. They may also carry divers or surface water rescue workers (Figure 22.5).

Figure 22.5 Fire Boat No.5, Hong Kong fire services department.

Chapter 23

Tugboats

A tugboat or tug is a marine vessel that manoeuvres other vessels by pushing or pulling them, with direct contact or a tow line. These boats typically tug ships that cannot move well on their own, such as those in crowded harbours or narrow canals, or those that cannot move at all, such as barges, disabled ships, log rafts, or oil platforms. Some are ocean-going, and some are icebreakers or salvage tugs. Early models had steam engines, and modern ones have diesel engines. Many have deluge guns, which help in firefighting, especially in harbours.

DEEP-SEA OR SEAGOING TUGS

Seagoing tugs (deep-sea tugs or ocean tugboats) fall into four basic categories. The standard seagoing tug with model bow, tows exclusively by way of a wire cable. In some rare cases, such as some US Navy fleet tugs, a synthetic rope hawser may be used for the tow in the belief that the line can be pulled aboard a disabled ship by the crew owing to its lightness compared with wire cable. The 'notch tug' can be secured by way of cables, or more commonly in recent times, synthetic lines that run from the stern to the tug to the stern of the barge. This configuration is used in inland waters where sea and swell are minimal because of the danger of parting the push wires. Often, this configuration is employed even without a 'notch' on the barge, but in those cases, it is preferable to have 'push knees' on the tug to stabilise its position. Model bow tugs employing this method of pushing always have a towing winch that can be used if sea conditions render pushing inadvisable. With this configuration, the barge being pushed might approach the size of a small ship, with the interaction of the water flow allowing a higher speed with a minimal increase in power required or fuel consumption.

The 'integral unit', or 'integrated tug and barge' (ITB), comprises specially designed vessels that lock together in such a rigid and strong method as to be certified as such by authorities (classification societies) such as the American Bureau of Shipping, Lloyd's Register of Shipping, the Indian Register of Shipping, Det Norske Veritas, and several others. These units

DOI: 10.1201/9781003342366-28

Figure 23.1 Abeille Bourbon.

stay combined under any sea conditions and the tugs usually have poor sea-keeping designs for navigation without their barges attached. Vessels in this category are legally considered to be ships rather than tugboats, and barges must be staffed accordingly. These vessels must show navigation lights compliant with those required of ships rather than those required of tugboats and vessels undertow.

'Articulated tug and barge' (ATB) units also use mechanical means to connect to their barges. The tug slips into a notch in the stern and is attached by a hinged connection, becoming an articulated vehicle. ATBs use Intercon and Bludworth connecting systems. ATBs are staffed as a large tugboat, having between seven and nine crew members. The typical American ATB displays navigational lights of a towing vessel pushing ahead, as described in the 1972 COLREGS (Figure 23.1).

HARBOUR TUGS

Compared with seagoing tugboats, harbour tugboats that are employed exclusively as ship assist vessels are smaller and their width-to-length ratio is often higher, due to the need for the tugs' wheelhouse to avoid contact with the hull of a ship, which may have a pronounced rake at the bow and stern. In some ports, there is a requirement for certain numbers and sizes

Figure 23.2 Tug *Michel Hamburg* assisting *COSCO Shipping Nebula.*

of tugboats for port operations with gas tankers. Also, in many ports, tankers are required to have tug escorts when transiting in harbours to render assistance in the event of mechanical failure. The port mandates a minimum horsepower or bollard pull, determined by the size of the escorted vessel. Most ports will have several tugs that are used for other purposes than ship assist, such as dredging operations, bunkering ships, transferring liquid products between berths, and cargo operations. These tugs may also be used for ship assist as needed. Modern ship assist tugs are 'tractor tugs' that employ azimuth stern drives (ASDs), propellers that can rotate 360 degrees without a rudder, or cycloidal drives (Figure 23.2).

RIVER TUGS

River tugs are also referred to as towboats or push boats. Their hull designs would make open ocean operations dangerous. River tugs usually do not have any significant hawser or winch. Their hulls feature a flat front or bow to line up with the rectangular stern of the barge, often with large pushing knees.

SALVAGE TUGS

A salvage tug, known historically as a wrecking tug, is a specialised type of tugboat that is used to rescue ships that are in distress or in danger of sinking or to salvage ships that have already sunk or run aground. Few tugboats have ever been truly fully dedicated to salvage work; most of the time, salvage tugs operate towing barges, platforms, ships, or performing other utility tugboat work. Tugs fitted out for salvage are found in small quantities around the world, with higher concentrations near areas with both heavy shipping traffic and hazardous weather conditions. Salvage tugs are used by specialised crew experienced in salvage operations (called salvors). The ships carry specialist equipment, including extensive towing provisions and extra tow lines and cables, with provisions for towing from both the bow and stern and at irregular angles, extra cranes, firefighting gear, deluge systems, hoses and nozzles, and a range of mechanical equipment such as common mechanical repair parts, compressed air gear, diving equipment, steel for hull patches, welding equipment, and pumps. Overall, total demand for salvage tug services has significantly decreased from its peaks in the years around World War II. The increasing sensitivity of societies and legal systems to environmental damage and the increasing size of ships has offset the decline in the number of salvage operations undertaken. Accidents such as major oil tanker groundings or sinkings may require extensive salvage efforts to try to minimise the environmental damage such as that caused by *the Exxon Valdez* (1989) oil spill, or the *Amoco Cadiz* (1978) and *Torrey Canyon* (1967) disasters (Figure 23.3).

Figure 23.3 Tugboat *Salvage Mark* in Jarvis Quay, Toronto, Ontario, Canada.

TENDERS

A ship's tender, usually referred to as a tender, is a boat, or a larger ship, used to service or support other boats or ships. This is done by transporting people or supplies to and from shore or another ship. A second and distinctly different meaning for tender is small boats carried by larger vessels, to be used either as lifeboats, as transport to shore, or both. For a variety of reasons, it is not always advisable to try to tie a ship up at a dock; the weather or the sea might be rough, the time alongside might be short, or the ship too large to safely fit. In such cases, tenders provide the link from ship to shore and may have a remarkably busy schedule of back-and-forth trips while the ship is in port. On cruise ships, lifeboat tenders fulfil double duties, serving as tenders in day-to-day activities, but fully equipped to function as lifeboats in an emergency. They are transported on davits just above the promenade deck and may appear to be regular lifeboats, but they are usually larger and better-equipped. Current lifeboat tender designs favour catamaran models since they are less likely to roll in calm to moderate conditions. They typically carry up to 100–150 passengers and two to three crew members. Before these ships were mass-produced, the main way to board a larger ship (ocean liners) was to board a passenger tender. Passenger tenders remained based at their ports of registry, and when a ship came through the area, the tender would tie up with the ship and embark passengers on an elevated walkway. These vessels were larger, had a greater passenger capacity, and a broader sense of individuality in their respective companies than the more modern tenders seen today. Because of their increased size, lifeboats and life preservers were commonplace on board these ships (with two boats being the standard amount for an average tender) (Figure 23.4).

Before the technologies that allow submarines and destroyers to operate independently matured by the latter half of the 20th century (and significantly during the World War II), they were heavily dependent upon tenders to perform most maintenance and supply. Their hull classification symbols in the US Navy were, respectively, AS and AD, while general repair ships were AR. Naval tenders fell out of use during the late 20th century, as the speed and range of warships increased (reducing the need for advanced basing). By the end of the 20th century, all the tenders in the US Navy had been inactivated except for two submarine tenders. Not completely willing to wean itself from tenders altogether – but with an eye towards reducing costs – the last two tenders remaining in active service have now been operationally turned over to the Military Sealift Command. Emory S. Land-class submarine tenders *USS Emory S. Land* and *USS Frank Cable* are now manned and operated by a 'hybrid' crew. The commanding officer and approximately 200 technicians are Naval personnel, while the operation of the ship is performed by civilian mariners. Prior to the turn-over, both ships had more than 1,000 sailors. While at this time the ships still bear the AS classification, the primary mission of both ships has been expanded well

Figure 23.4 Havila Commander, departing Aberdeen Harbour, Scotland.

beyond submarines to include service and support of any Naval vessel in their operational area. Under the traditional Navy classification, both ships should be reclassified as AR (auxiliary repair); however, since now operated by the MSC, it is doubtful that such a reassignment will occur. *USS Emory S. Land* is forward deployed in the Indian Ocean at Diego Garcia, while *USS Frank Cable* is forward deployed in the Pacific at Polaris Point, Apra Harbour, Guam. Such forward deployments are designed to provide service and support at the very great distances of the Western Pacific.

Two tenders, *SS Nomadic* and *SS Traffic*, were built for the White Star Line by Harland and Wolff to serve the liners *RMS Olympic* and *RMS Titanic* at Cherbourg. Nomadic survives as a museum ship and is the last remaining vessel built for the White Star Line still in existence.

PROPULSION SYSTEMS

The first tugboat, *Charlotte Dundas*, was built by William Symington in 1801. She had a steam engine and paddle wheels and was used on the rivers of Scotland. Paddle tugs proliferated thereafter and were a common sight for a century. In the 1870s schooner, hulls were converted to screw tugs, and fitted with compound steam engines and scotch boilers, providing as much as 300 bhp. Within a few decades, steam tugs were used in every harbour around the world for towing and ship berthing. Tugboat diesel engines typically produce between 500 and 2,500 kW (~ 680–3,400 bhp), but larger boats (used in deep waters) can have power ratings of up to 20,000 kW (~ 27,200 hp). Tugboats usually have an extreme power to tonnage ratio; normal cargo and passenger ships have a power to tonnage radio [in Kw: gross registered tonnage (GRT) of 0.35–1.20], whereas large tugs typically are 2.20–4.50 and small harbour-tugs are 4.0–9.5. The engines are often the same as those used in railroad locomotives but typically drive the propeller mechanically instead of converting the engine output to power electric motors, as is common for diesel-electric locomotives. For safety, tugboat engines often feature two of each critical part for redundancy. A tugboat is typically rated by its engine's power output and its overall bollard pull. The largest commercial harbour tugboats in the 2000s–2010s, used for towing container ships or similar, had around 60–65 short tonnes-force (530–580 kN) of bollard pull, which is described as fifteen short tonnes-force (130 kN) above 'normal' tugboats. Tugboats are highly manoeuvrable, and various propulsion systems have been developed to increase manoeuvrability and increase safety. The earliest tugs were fitted with paddle wheels, but these were soon replaced by propeller-driven tugs. Kort nozzles (see below) have been added to increase thrust-to-power ratio. This was followed by the nozzle-rudder, which omitted the need for a conventional rudder. The cycloidal propeller (see below) was developed prior to World War II and was occasionally used in tugs because of its manoeuvrability. After World War II, it was also linked to safety due to the development of the Voith Water Tractor, a tugboat configuration that could not be pulled over by its tow. In the late 1950s, the Z-drive or (azimuth thruster) was developed. Although sometimes referred to as the Aquamaster or Schottel system, many brands exist, including Steerprop, Wärtsilä, Berg Propulsion, and so forth. These propulsion systems are used on tugboats designed for tasks such as ship docking and marine construction, whereas conventional propeller and rudder configurations are more efficient for port-to-port towing.

Kort nozzle

The Kort nozzle is a sturdy cylindrical structure around a special propeller having minimum clearance between the propeller blades and the inner wall of the Kort nozzle. The thrust-to-power ratio is enhanced because the water approaches the propeller in a linear configuration and exits the nozzle the same way. The Kort nozzle is named after its inventor (Ludwig Kort, 1934), but many brands now exist.

Cyclorotor

The cycloidal propeller is a circular plate mounted on the underside of the hull, rotating around a vertical axis with a circular array of vertical blades (in the shape of hydrofoils) that protrude out of the bottom of the ship. Each blade can rotate itself around a vertical axis. The internal mechanism changes the angle of attack of the blades coordinated with the rotation of the plate, so that each blade can provide thrust in any direction, like the collective pitch control and cyclic in a helicopter.

Carousel

A recent Dutch innovation is the carousel tug, winner of the Maritime Innovation Award at the Dutch Maritime Innovation Awards Gala in 2006. It adds a pair of interlocking rings to the body of the tug, the inner on the boat, the outer on the ship by winch or towing hook. Since the towing point rotates freely, the tug is exceedingly difficult to capsize.

In this penultimate chapter, we have looked at some of the main types and the defining features of tugboats. In the next and final chapter, we will discuss the role and function of the Royal Fleet Auxiliary (RFA). The RFA is an unusual creation in that it is a fleet of merchant-type vessels (ROROs, support vessels, and fleet tankers), which are manned by civilian seafarers but sail under the authority of the Royal Navy.

Part V

Royal Fleet Auxiliary

Chapter 24

Royal Fleet Auxiliary

The Royal Fleet Auxiliary (RFA) is the naval auxiliary fleet owned by the UK's Ministry of Defence. The RFA provides vital logistical and operational support to the Royal Navy and Royal Marines by ensuring the Royal Navy is supplied and supported by providing fuel and stores through replenishment at sea (RAS), transporting Royal Marines and British Army personnel around the world, providing medical care, and transporting equipment and materiel wherever British Forces are active. In addition, the RFA acts independently providing humanitarian aid, counter-piracy, and counter-narcotic patrols together with assisting the Royal Navy in preventing conflict and securing international trade. The RFA is a uniformed civilian branch of the Royal Navy. This means that the RFA personnel are civilian employees of the Ministry of Defence and members of the Royal Naval Reserve and Sponsored Reserves. Although RFA personnel wear Merchant Navy rank insignia on their uniforms, they are classed as part of the Naval Service. RFA vessels are commanded and crewed by RFA officers and ratings, just as any other merchant vessel, but are augmented with regular and reserve Royal Navy personnel who perform specialised functions such as operating and maintaining helicopters or providing hospital facilities. Royal Navy personnel are also needed to operate the shipboard weapons, such as the Phalanx; however, a certain number of defensive weapons (such as the Bushmaster 30mm cannon) are operated by RFA personnel. It is worth mentioning that although the RFA is a civilian component of the Royal Navy, the fact that RFA vessels are armed (defensively) means that they are recognised as combatant ships under the Geneva Convention. This means that RFA crews are not afforded the same protections as purely civil merchant vessels.

The RFA was first established in 1905 to provide coaling ships for the Royal Navy in an era when the change from sail to coal-fired steam engines as the main means of propulsion meant that a network of bases around the world with coaling facilities or a fleet of ships able to supply coal was necessary for the British Fleet to operate away from its home country. Since the Royal Navy of that era possessed the largest network of bases around the world, initially, the RFA took a minor role. It was not until World War II that the Royal Navy recognised the true worth of the RFA, as British

DOI: 10.1201/9781003342366-30

warships were often far from available bases, either due to the enemy capturing British or Allied bases or as is the case in the Pacific, due to the sheer distances involved. World War II also saw naval ships staying at sea for much longer periods than had been the case since the days of sail. During this time, techniques for the RAS were developed. The auxiliary fleet comprised a diverse collection, with not only RFA ships, but also commissioned warships and merchantmen as well. After 1945, the RFA became the Royal Navy's main source of support in the many conflicts that the Royal Navy was involved in. The RFA performed important service to the Far East Fleet off Korea from 1950 until 1953, when sustained carrier operations were mounted in Pacific waters. During the extended operations of the Konfrontasi in the 1960s, the RFA was also heavily involved. As the network of British bases overseas shrank during the years when the British Empire diminished, the Royal Navy became increasingly reliant on the RFA to supply its ships during routine deployments. The RFA played an integral role in the largest naval war since 1945, the Falklands War in 1982 between Britain and Argentina. In that conflict, one RFA vessel was lost, and another was irreparably damaged. The RFA has received further battle orders from service in the first Gulf War (1990–1991), the Kosovo War (1998–1999), the Afghanistan Campaign (2001–2021), the invasion of Iraq (2003–2011), and latterly, the civil war in Syria (2011–present). In July 2008, the RFA was presented with a Queen's Colour, an honour unique to a civilian organisation (Figure 24.1).

Figure 24.1 *RFA Argus* in the Caribbean Sea – Summer 2020.

RFA FLEET

Ships in RFA service carry the ship prefix RFA, and fly the Blue Ensign defaced with an upright gold killick anchor. All Royal Fleet Auxiliaries are built and maintained to Lloyd's Register and Department for Transport standards. The most significant role provided by the RFA is RAS; therefore, the mainstay of the current RFA fleet is the replenishment class of ships. The *Wave class* are 'Fleet Tankers', which not only primarily provide underway refuelling to Royal Navy ships but can also provide a limited amount of dry cargo. The *Tide class* are 'Fast Fleet Tankers', which were ordered in February 2012 and delivered in 2017. From 2022, only the *Tide class* are expected to remain active with both *Wave class* vessels being placed in extended readiness (uncrewed reserve). The Fort Victoria is a 'one-stop' replenishment ship, capable of providing underway refuelling and dry cargoes (i.e. rearming, victualling, and spares). The older *Fort Rosalie class* ships provided only dry cargoes. Both *Fort Rosalie class* vessels were placed in reduced (base maintenance period) or 'extended readiness' (uncrewed reserve) in June 2020. The 2015 Strategic Defence and Security Review stated that three new 'Fleet Solid Support' Ships were to be built and bidding for the contract was to start in late 2016. In 2019, this competition was stopped in the face of criticism that the competition permitted bids to build the ships from outside the United Kingdom. In May 2020, then Defence Secretary Ben Wallace stated that the competition was likely to restart in September 2020; however, the start was then delayed to the 'spring' of 2021. The 2021 Defence White Paper confirmed that both *Fort Rosalie class* ships would be decommissioned and eventually replaced by new Fleet Solid Stores Support Vessels. In October 2021, both ships were sold to Egypt (Figure 24.2).

The *Wave class*, *Tide class*, and *Fort Victoria* incorporate aviation facilities, providing aviation support and training facilities as well as vertical replenishment capabilities. They are capable of operating and supporting Merlin and Lynx Wildcat helicopters, both of which are significant weapons platforms. The presence of aviation facilities on RFA ships allows for them to be used as 'force multipliers' for the task groups they support in line with Royal Navy doctrine. The RFA is tasked with the role of supporting Royal Navy amphibious operations through its three *Bay class* dock landing ships (LSD). Typically, one *Bay class* is also assigned as a permanent 'mothership' for Royal Navy mine countermeasures vessels in the Persian Gulf. The 2021 Defence White Paper proposed the acquisition of a new class of up to six Multi-Role Support Ships to support littoral strike operations. These seemed likely to replace the *Bay class* ships by the 2030s. In the interim, the White Paper proposed to upgrade one of the *Bay class* vessels with permanent hangar facilities to conduct the littoral strike role. The unique support ship in the fleet is the aviation training ship *Argus*, a converted RORO ship. She is tasked with peacetime aviation training and support. On active operations, she becomes the primary casualty receiving ship (PCRS), a hospital ship. She cannot be described as such and is not afforded such protection under the

Figure 24.2 RFA Tidespring.

Geneva Convention, as she is armed. She can, however, venture into waters too dangerous for a normal hospital ship. *Argus* completed a refit in May 2007 intended to extend her operational life to 2020. As of 2021, *Argus* was still in service but expected to retire from service in 2024. The 2021 Defence White Paper did not specifically mention her replacement. However, her functions are likely eventually to be taken over by the new Fleet Solid Stores Support ships approved for acquisition in the 2021 Defence White Paper.

The *Point class* sealift ships were acquired in 2002 under a £1.25 billion ($156 billion) private finance initiative with Foreland Shipping known as the 'Strategic Sealift Service'. These ships are civilian merchant navy vessels leased to the Ministry of Defence as and when needed. Originally, six ships were part of the deal, allowing the Ministry of Defence the use of four of the ships with two being made available for commercial charter. These latter two were released from the contract in 2012. The Ministry of Defence also contracts to secure fuel supplies for facilities overseas. This requirement was maintained through the charter of the vessel *Mærsk Rapier*. The ship was tasked with supplying fuel to the UK's various naval establishments at home and overseas, as well as providing aviation fuel to RAF stations at Cyprus, Ascension Island, and the Falklands. The Ministry of Defence chartered the vessel to commercial companies during periods when she was not in use for defence purposes. Since the end of the contract for the use of *Mærsk Rapier*, a further contract for the use of another tanker, renamed the *Raleigh Fisher*, was secured (Figure 24.3).

As of 2022, there are 11 ships in service with the RFA with a total displacement of approximately 329,000 tonnes. These figures exclude merchant navy vessels under charter to the Ministry of Defence (Table 24.1).

Figure 24.3 *RFA Fort George* at the jetty of the naval fuel depot, mouth of Loch Striven (near Port Lamont), Scotland.

Table 24.1 RFA fleet (as of 2022)

Replenishment					
Class	*Ship*	*Pennant*	*Entered service*	*Displacement*	*Type*
Tide Class	*RFA Tidespring*	A136	2017	39,000 metric tonnes	Replenishment Tanker
	FRA Tiderace	A137	2018		
	RFA Tidesurge	A138	2019		
	RFA Tideforce	A139	2019		
Wave Class	*RFA Wave Knight*	A389	2003	31,500 metric tonnes	Fast Fleet Tanker
	RFA Wave Ruler	A390	2003		
Fort Victoria Class	*RFA Fort Victoria*	A387	1994	33,675 metric tonnes	Multirole Replenishment
Dock Landing					
Class	*Ship*	*Pennant*	*Entered Service*	*Displacement*	*Type*
Bay Class	*RFA Lyme Bay*	L3007	2007	16,160 metric tonnes	Dock Landing Ship Auxiliary
	RFA Mounts Bay	L3008	2006		
	RFA Cardigan Bay	L3009	2006		

(Continued)

Table 24.1 (*Continued*) RFA fleet (as of 2022)

Aviation support/Casualty evacuation					
Class	*Ship*	*Pennant*	*Entered service*	*Displacement*	*Type*
	RFA Argus	A135	1988	28,081 metric tonnes	Aviation training and primary casualty receiving ship

Ministry of Defence Sealift / Support Vessels					
Class	*Ship*	*Owner*	*Entered Service*	*Displacement*	*Type*
Point Class	*MV Hurst Point*	Foreland Shipping	2002	23,000 metric tonnes	RORO Sealift
	MV Eddystone		2002		
	MV Hartland Point		2002		
	MV Anvil Point		2003		
	MV Raleigh Fisher	James Fisher & Sons	2005	35,000 metric tonnes	Tanker

Index

F

G

H

I

N

O